Outsourced Empire

Outsourced Empire

How Militias, Mercenaries, and Contractors Support US Statecraft

Andrew Thomson

First published 2018 by Pluto Press
345 Archway Road, London N6 5AA

www.plutobooks.com

British Library Cataloguing in Publication Data
A catalogue record for this book is available from the British Library

ISBN   978 0 7453 3705 0   Hardback
ISBN   978 0 7453 3703 6   Paperback
ISBN   978 1 7868 0262 0   PDF eBook
ISBN   978 1 7868 0264 4   Kindle eBook
ISBN   978 1 7868 0263 7   EPUB eBook

This book is printed on paper suitable for recycling and made from fully
managed and sustained forest sources. Logging, pulping and manufacturing
processes are expected to conform to the environmental standards of the
country of origin.

Typeset by Stanford DTP Services, Northampton, England

Simultaneously printed in the United Kingdom and United States of America

Contents

Abbreviations and Acronyms

AA	Air America
AAA	American Anti-Communist Alliance
ALP	Afghan Local Police
AMISOM	African Union Mission to Somalia
ANA	Afghan National Army
ANAP	Afghan National Auxiliary Police
ANMC	American National Management Corporation
ANP	Afghan National Police
ANSENSAL	*Agencia Nacional de Seguridad de El Salvador*—National Security Agency of El Salvador
ARVN	Army of the Republic of Vietnam
AUC	*Auto-Defensas Unidas de Colombia*—United Self-Defense Forces of Colombia
CAD	*Comités de Auto-Defensa* (Peru)
CAFGU	Civilian Armed Force Geographical Unit (Philippines)
CAT	Civil Air Transport
CEA	California Eastern Airways
CIA	Central Intelligence Agency
CIDF	Civilian Irregular Defense Forces
CIDG	Civilian Irregular Defense Groups
CMA	Civilian Military Assistance
CORDS	Civil Operations and Revolutionary Developmental Support
CORU	Coordination of United Revolutionary Organizations
CPA	Coalition Provisional Authority (Iraq)
CSDF	Civilian Self-Defense Forces
DDR	Disarmament, Demobilization, and Reintegration
DIA	Defense Intelligence Agency
DoD	Department of Defense
EAST	Eagle Aviation Services and Technology
ELN	*Ejército de Liberación Nacional*—Army of National Liberation (Colombia)
ERU	Emergency Response Unit (Iraq)

EZLN	*Ejercito Zapatista de Liberacion Nacional*—Zapatista Army of National Liberation (Mexico)
FARC	*Fuerzas Armadas Revolucionarias de Colombia*—Revolutionary Armed Forces of Colombia
FATA	Federally Administered Tribal Areas
FDI	Foreign Direct Investment
FID	Foreign Internal Defense
FM	Field Manual
FSA	Free Syrian Army
GNA	Government of National Accord (Libya)
HN	Host Nation
IDAD	Internal Defense and Development
IMET	International Military Education and Training
IMF	International Monetary Fund
ISIS	Islamic State
ISOF	Iraqi Special Operations Forces
ITT	International Telephone and Telegraph
IW	Irregular Warfare
JCET	Joint Combined Exercise Training program
LIC	Low-Intensity Conflict
LOGCAP	Logistics Civil Augmentation Program
MACV	Military Assistance Command, Vietnam
MAS	*Muerte a los Secuestradores*—Death to Kidnappers (Colombia)
MILF	Moro Islamic Liberation Front (Philippines)
MOOTW	Military Operations Other Than War
MNCs	Multinational Corporations
MPRI	Military Professional Resources Incorporated
MTTs	Mobile Training Teams
NAFTA	North American Free Trade Agreement
NATO	North Atlantic Treaty Organization
NLF	National Liberation Front
NPA	New People's Army (Philippines)
NSAM	National Security Action Memorandum
NSC	National Security Council
OIDP	Overseas Internal Defense Policy
OMB	Office of Management and Budget
ORDEN	*Organización Democratica Nacionalista*—Nationalist Democratic Organization (Nicaragua)

PAC	*Patrulla de Auto-Defensa Civil*—Civil Self-Defense Patrol
PGMD	Pro-Government Militias Database
PMCs	Private Military Companies
PNAC	Project for a New American Century
PRD	*Partido de la Revolución* (Mexico)
PRI	*Partido Revolucionario Institucional*—Institutional Revolutionary Party (Mexico)
PRU	Provincial Reconnaissance Unit
PSYOPS	Psychological Operations
RMA	Revolution in Military Affairs
SAIC	Science Applications International Corporation
SANG	Saudi Arabian National Guard
SAS	Special Air Service
SAT	Southern Air Transport
SDF	Syrian Democratic Forces
SIJIN	*Sección de Investigaciones Judiciales e Inteligencia de la Policía*—Judicial Police Investigative and Intelligence Unit (Colombia)
SOA	School of the Americas
SOCOM	United States Special Operations Command
SOUTHCOM	United States Southern Command
TWA	Trans World Airlines
UAV	Unmanned Aerial Vehicle
UCLA	Unilaterally Controlled Latino Asset
UK	United Kingdom (of Great Britain)
UNITA	*União Nacional para a Independência Total de Angola*—National Unity for Total Independence of Angola (Angola)
US	United States (of America)
USAID	United States Agency for International Development
VC	Viet Cong
WACL	World Anti-Communist League
WTO	World Trade Organization
YPG	*Yekîneyên Parastina Gel*—Kurdish People's Protection Units (Syria)

Introduction

Before the US-led Coalition invaded Iraq to topple Saddam Hussein in March 2003, CIA and Special Forces teams had already entered the country. Their mission was to mobilize internal opposition to weaken Hussein's regime from within and to prepare the ground for the arrival of US military forces. In February 2002, joint CIA and Special Force teams (labeled Northern Iraq Liaison Elements) met with Kurdish Peshmerga and organized for their fighters to wage an insurgency against the government. The CIA trained, paid, and coordinated Peshmerga insurgents to conduct raids, ambushes, and "sabotage" missions.[1] Kurdish militias also helped identify targets for US aerial strikes. During the US-led military invasion (March–May 2003), these Kurdish forces operated alongside US troops to take over key areas in the north of the country and to push Saddam's army to the south. In the months leading to the invasion, the CIA also spent millions of dollars developing a guerrilla paramilitary unit called the "Scorpions," with the aim of attacking infrastructure such as bridges and government facilities and to cultivate support amongst the population for a rebellion against the Iraqi state.[2] In addition, US officials approved funding for organizing Iraqi exiles into militias to help in the invasion.[3] The Department of Defense (DoD) trained a semi-official military force of Iraqis in Hungary, for example, to enhance US military-civil relations in Iraq during the invasion.[4] Many non-state armed groups aided the US military before and during the invasion of Iraq.

After the invasion and the removal of the Baathist party from power, the US struggled to contain a growing multi-dimensional insurgency that featured anti-occupational opposition, remnants of the Baathist party, and Islamist Jihadists. As part of the ensuing counterinsurgent project to stabilize a new political and economic order in Iraq, US military forces were reinforced by a variety of non-state armed actors. Private military companies (PMCs) and private security contractors flooded onto the scene. Most operated in support roles, such as base maintenance, allowing the US military to concentrate on its front-line tasks. Many others formed central parts of the coercive mechanisms of US statecraft, conducting interrogations, operating technical military equipment, and serving as security to protect vital infrastructure, among many other roles. Private military contractors were also hired to train personnel for the new Iraqi military and police services as well as select counter-insurgent militia forces. Meanwhile, US military planners supported militia and commando paramilitary forces. Many observers and US personnel alike compared these forces to US support for death squads in Latin America during the Cold War and paramilitary programs implemented in Vietnam, such as the Phoenix program.[5] Even the success of the US troop "surge," which saw tens of thousands of US soldiers enter the country to overpower the insurgency, partially hinged on turning former insurgents and gaining the support of Sunni tribal groups in a militia program known as the "Sons of Iraq."[6]

As more power was transferred to the newly established Iraqi government, similar patterns of military public-private partnerships continued, with US advisors and the Iraqi state employing PMCs, paying mercenaries and warlord factions, and cajoling a myriad of militia forces.[7] Towards the end of the US-led Operation Iraqi Freedom, the US adopted a "train, advise, and assist" mission to support the Iraqi government and its military and police forces without the involvement of significant numbers of US soldiers on the ground. In this advisory role, US officials acquiesced to and forged alliances with militias and paramilitary forces as a counterinsurgent vanguard.[8] US assistance to the Iraqi Armed Forces was also leveraged through PMCs and security contractors paid for by both the US and the Iraqi state. Then, when the Islamic State (ISIS) invaded some of the northern territories of Iraq, expanding their territorial acquisitions across the Syrian-Iraq border, the US and its Iraqi state military allies supported a collection of militias to defeat ISIS and reassert control over ISIS-held regions. The US, for its part, assisted and trained factions of the Popular Mobilization Forces, an umbrella Shi'ite militia group, Kurdish militia forces, and a diverse range of other militias in Syria to militarily defeat ISIS, driving it from the region.[9] Alongside US military aerial strikes and the US-backed Iraqi Armed Forces, US strategies towards the region have depended on applications of force that allowed the US to distance itself from its interventionist policies. By December 2017, a coordinated campaign involving US and Iraqi military forces, PMCs, and US-backed Iraqi militias had retaken key cities such as Mosul, bringing them back under Iraqi government control.

The US-led invasion, the Coalition occupation of Iraq, and US strategies after handing control over to an Iraqi government were reinforced at every stage by a collection of militias, local armed groups, mercenary-like forces, and hired PMCs. This constituted part of an outsourced means of pursuing US foreign policy objectives. Broadly speaking, US objectives in Iraq not only included the removal of an autocratic leader and the preemptive annihilation of an alleged international security threat to the US and its Western allies, but also aimed at the reorientation of the Iraqi political economy so as to make it more germane to a US-led global order. This entailed a transnationalization of the Iraqi economy, in line with neoliberal economic models.[10] This process, and the entire US-led project to bring Iraq to heel and re-engineer its political economy, depended on configurations of para-institutional military forces that extended beyond US or Iraqi state structures and official military departments. The military and security dimensions of the US-engineered transformation of the Iraqi political economy and the stabilization of a new government included complexes of non-state coercive forces. This was tantamount to an outsourcing of imperialism (or at least parts of it). Effectively, this meant that public/private and global/local forces aided in the "governance" of Iraq.

The intervention into Iraq is not the only example of such trends in US foreign policy. US advisors have emphasized working "with or through" such "surrogate" forces around the world towards similar objectives. US counterinsurgents in Afghanistan co-opted warlord factions and mobilized civilians to fight alongside official Afghan military and police forces to pacify a Taliban-Haqqani and anti-occupational insurgency towards the stabilization of a specific order conducive to US and transnational capital interests. PMCs and security contractors have significantly multiplied these efforts. By mid-2017, PMCs in Afghanistan outnumbered US troops three to one, as President Trump mulled over proposals to hire more.[11] Such forces overlapped and were often coordinated together, forming complexes of para-institutional power in US statecraft. US planners hired PMCs to train militia forces, and in other cases PMCs further subcontracted their work to local mercenary and warlord factions.[12] In short, global/local and public/private patterns and relationships have been central to US-led efforts to stabilize a post-Taliban order. Similar arrangements were set in motion across the Middle East and around the world, in conflict hotspots such as Libya, Syria, and Yemen as well as in Colombia and the Philippines, among others. Policymakers in Washington and their foreign allied state partners have presided over extensive public/private partnerships for the policing of peripheral areas around the world.

The US delegated central tasks to militias, mercenaries, PMCs, and a variety of similar non-state armed actors throughout the "War on Terror" and has continued to do so up until the present day. Washington has ensured that networks of para-institutional forces bear the brunt of its counterinsurgency and unconventional wars. US policymakers and military planners have helped to mobilize local collaborators and worked with "private" armies to contribute towards the internal defense of allied states, and, in many other cases, to weaken and ultimately overthrow unwanted regimes. This forms a pattern in which US military strategists have argued, in their words, for a "light footprint" approach to US engagements abroad, sending CIA and Special Forces to work "with or through" a collection of "irregular forces" as "force multipliers" to "extend US military reach" around the world.[13] As US military bases, ships, aircraft, and personnel are stretched across the planet with defense expenditures exceeding 40 percent of the world's total and providing military training to most of the world's countries, alliances with militias, mercenaries, contractors, and other forces have buttressed US-supported counterinsurgency and unconventional warfare.

Outsourcing, Privatization, and Alliances with Para-Institutional Military Forces in US Foreign Policy

These developments have prompted substantial interest in the outsourcing of US foreign policy to PMCs, the delegation of tasks to militias and paramilitaries, and proxy warfare.[14] Underpinning the majority of these existing studies, though, is a propensity to exclusively focus on specific types of parallel military forces, such as PMCs, and to emphasize each type of actor's relative newness as a product of a post-Cold War security environment. Experts on US privatization to PMCs, for example, tend to agree that shifts in the international environment and US (and other core states') military downsizing at the end of the Cold War led to the emergence of a privatized military industry, and that this represented a "profound shift" or "revolution" in US military power projection.[15] Similarly, many commentators have placed an examination of the US use of "surrogates" and "irregular" forces in the context of Kaldor's "new wars," in which civil wars have ostensibly experienced an increase in prevalence of non-state actors in the post-Cold War era, and related these developments to "fourth-generation warfare" or "hybrid wars," which similarly postulate significant changes in the ways the US (and other actors) conduct military affairs.[16] In short, there has been a tendency to depict US outsourcing, privatization to PMCs, and alliances with local militia groups as a relatively new phenomenon after the Cold War and as developing in processes detached from US foreign policy itself. The general line of reasoning has been that the US, in delegating roles to such forces, is simply adapting to the existing global "market for force" or updating its strategies to "modern ways of war." There has been little consideration of militias, mercenaries, PMCs, and other actors together, and how and why they collectively form part of broader processes at play in US foreign policy.

This book examines how and why militias, mercenaries, and contractors have collectively been prominent in US foreign policy throughout the post-WWII era. The main argument is that we can better understand the workings and dynamics of US outsourcing, privatization, and alliances with a diverse range of non-state armed actors through an appreciation of how these practices developed in the context of US imperialism. This book examines the development of outsourcing in US foreign policy and how and why "irregular" para-institutional armed actors have been increasingly central to US efforts to create and maintain a US-led liberalized global order throughout the post-WWII era. It traces the evolution and growth of a nexus, or the composite relationships between the US military and a collection of para-institutional forces in the conduct of coercive modes of statecraft, namely counterinsurgency and unconventional warfare. US imperialism has been exercised, in US and US-supported counterinsurgency and unconventional warfare, through the development of *para-institutional complexes*—direct and indirect connections with a variety of parallel armed actors, such as localized militias, mercenaries, insurgents, paramilitaries, and global or international PMCs and others.

This focus necessarily concentrates on the military dimensions of US imperialism. Building on existing accounts of US foreign policy and its relationship to processes of globalization, this book posits that throughout the post-WWII era the US has occupied a key managerial role in the international arena by working to re-orient states towards the fluid functioning of global capitalism.[17] The US has led the way in transforming state arrangements, political orders, and corresponding sets of political and economic policies in countries primarily in the South so that they are conducive to the requirements of an expanding interconnected and global capitalist system. While the US has employed various means and methods towards this, in locations where substantial threats to the prevailing order exist, counterinsurgency and unconventional warfare have typically constituted the primary modes of politico-military coercive statecraft towards advancing and protecting transnational elite interests and those of multi-national corporations (MNCs).[18] Counterinsurgency and unconventional warfare have been deployed either through direct US intervention or more indirectly via military assistance to allied ruling strata in foreign countries to "stabilize" favorable political and economic arrangements and policies abroad, principally for access to strategic raw materials, new markets, and stable trading and investment climates. Para-institutional actors of various types have, in turn, been central to these processes of "stabilization" in US statecraft and to US global military presence. This has produced different modalities of imperial coercive power as the US and allied state formations work "with and through" various para-institutional agents.

This book also traces the evolution and the development of US use of coercive para-institutional forces in these forms of statecraft. This historical component focuses on the increasing reliance of various military-paramilitary relationships in the prosecution of US statecraft since the ascendancy of US power in the post-WWII international system. Against the backdrop of broader continuities in US imperialism, this book maps out the trends and events throughout the post-war period that have influenced the development of patterns of delegation of force and specific shapes it has taken. It demonstrates that contemporary sets of military-paramilitary relationships as well as practices of military outsourcing in US foreign policy have their origins within US strategies during the Cold War. The Vietnam War and the subsequent American domestic political costs associated with intervention encapsulated in the "Vietnam syndrome" helped perpetuate para-institutional force relations and have since become consolidated in US foreign policy in the "War on Terror." This historical examination takes into account and incorporates the advancement of capitalism and its neoliberal forms that surfaced towards the end of the 1980s and early 1990s as well as how transnational interests and MNCs' actions have contributed to increases in the propensity to form alliances with various para-institutional forces. This helps to highlight how contemporary practices of outsourcing, privatization, and alliances with various para-institutional armed forces have a longer history in US foreign policy than is commonly portrayed. Such practices have their roots in Cold War interventions, and the main precursors to contemporary PMCs in US foreign policy formed from the need for covert logistical and aerial combat capacities during such interventions. Practices of outsourcing and alliances with various paramilitary actors in US foreign policy have developed within the context of US imperialism.

The inclusion of para-institutional forces and the ways in which global and local non-state coercive agents fit into broader patterns and practices of US statecraft also helps to shift our understanding of US imperialism. Many scholars and observers concentrate on the US state, principally its military forces, as the primary tool of an American imperialism. For many observers, the sheer scale of US military power and its global presence defines an American informal "empire." Analysts concentrate on issues such as US militarism,[19] how US military bases abroad afford the US the ability to strike anywhere, anytime,[20] and emphasize US supremacy at sea, on land, and even in space (and the related concept of "full spectrum dominance").[21] Others examine US armed forces' military superiority and ability to conduct counterinsurgency,[22] or highlight US military training to allied countries and how foreign states fulfill the tasks of US foreign policy endeavors.[23] By incorporating an understanding of how coercive para-institutional complexes have been central to US counterinsurgency and unconventional warfare and how these modes of statecraft operate towards informal imperialism, this analysis demonstrates how US power is often channeled through or in conjunction with non-state or surrogate forces and collaborators. This represents a step towards a better understanding of how "irregular" forces have been instrumental to US strategies designed to influence political and economic conditions abroad. The US's ability to influence the politics and economics of peripheral states, principally in the global South, is actualized through composite relations and collaborations "beyond" the formal institutions of both the US and allied states. Instead, it has relied to an increasing extent on outsourcing, "private" armies, and collaborations with local agents. The US could not enforce a semblance of a US-led liberalized global order or an informal "empire" on its own, or solely via its official armed forces, and thus has developed patterns of relationships with various para-institutional military forces towards the "stabilization" of beneficial political and economic arrangements in countries in the South. Rather than the US military and other agencies, such as the CIA, the burdens of imperial policing have increasingly fallen to a pliable or flexible coercive workforce of indigenous collaborators, militias, and "private" armies. In bringing together an understanding of US imperialism, its relationship to global capitalism, and an examination of para-institutional complexes in US statecraft, this book highlights processes in which imperialism is exercised and configured through various means.

Of course, many imperial, and in particular, colonial, powers have delegated coercive tasks to non-state armed assets and local collaborators. Outsourcing in this manner is not unique to the US. Many observers have recorded and examined European colonial conquest and the networks of collaborators that made it possible.[24] For example, the UK has a long historical record, especially during the height of the British Empire, of delegating its imperial violence and policing roles not only to native troops but also to mercenaries, pirates (privateering), and other forces.[25] This also included local armies, regiments, and conscripts often organized into semi-official and official colonial military institutions, such as the Ghurkas.[26] Very rarely, if ever, in modern history have colonial powers dominated the periphery through the armed forces of state structures alone, without the aid of "private" armies. Harnessing local collaboration and leveraging "private" armies has been a fairly standard way imperial states have conquered, controlled, and policed far-away peripheral colonies.

Similarly, many countries currently delegate military functions to para-institutional forces such as militias, mercenaries, and PMCs. The UK and France, for example, have frequently employed mercenaries and PMCs and depended significantly on local non-state armed forces.[27] British covert wars in Yemen in 1962–67 operated with or through mercenary and paramilitary forces towards stabilization goals in a fashion similar to the processes in US imperialism under investigation in this book.[28] And, of course, the Soviets and Cubans backed numerous subversive forces in the Third World during the Cold War. Other states have used proxy forms of engagement for more limited foreign policy goals. For example, Libyan leader Muammar Gaddafi supported a variety of insurgent organizations internationally and propped up his regime internally via supporting local militias, informants, and power brokers. Western media outlets frequently remind us that Iran is also deeply involved in what is deemed to be proxy intervention in Afghanistan, Iraq, and Syria, asserting its foreign policy objectives through local paramilitary forces.

Practices of privatization, outsourcing, and alliances with various local paramilitary forces are certainly not unique to the US and instead represent fairly common ways for states to project power internationally. In this sense, Hughes is correct in asserting that more critically minded scholars who

> ... indict the USA and other Western countries for employing what they regard as a form of indirect aggression which breaches international norms of international behavior ... are under an intellectual obligation to acknowledge that there are few states within the international system which can claim innocence of such activity.[29]

So, what makes the US of particular interest?

How US coercive imperial statecraft has been extended by para-institutional mechanisms is worthy of our attention for a number of reasons. First and foremost, the US is the world's greatest superpower, and the sheer scale of a military-paramilitary nexus including PMC operations and connections to various non-state actors via its myriad military relations worldwide is unsurpassed by any other country. Para-institutional forces significantly bolster and support the global reach and influence of the US military. Secondly, while other countries may have used para-extensional means to complete limited objectives abroad, para-institutional agents have been central to US imperialism and by extension to both the processes and architecture that binds the contemporary US-led liberalized global order. Para-institutional complexes of intervention have been central to policing of the periphery and myriad sets of relationships between state and para-institutional forces are embedded within deeply political settings and bound up in processes of global capital expansion and accumulation. Finally, this has wider significance for the stabilization of specific state formations throughout much of the global South. US statecraft and counterinsurgency and unconventional warfare interventions and the para-institutional relations that have accompanied them have had considerable influence over developmental pathways of countries in the global South. Collections of non-state armed actors have aided in the creation and maintenance of a US-led liberalized global order and in the advancement of particular corresponding political and economic policies abroad. What makes an analysis of outsourcing in US imperialism so important, then, is not the mere fact that the US uses PMCs and various paramilitary forces to multiply or project US military power, but the scope of it and the centrality of these forces in US imperialism and "stabilization" of certain political and economic arrangements in countries throughout the global South. Para-institutional forces in US foreign policy are unique in the specific ways they contribute to processes of the formation of the contemporary liberalized global order. It is a central aspect of US imperial relations and has helped to underpin US-led globalization.

Conceptualizing Para-Institutional Forces

The term "para-institutional forces" refers to parallel armed actors that operate outside conventional or official chains of command, such as militias, death squads, paramilitary groups, civilian defense forces, insurgents, mercenaries, contractors, PMCs, and other irregular groups. This term draws from the rich literature on Latin American civil wars in which para-institutional actors were seen as "loosely—and usually covertly—affiliated with organs of the state, that may depend on them for support and that may even have been created or licensed by the state itself to collaborate in the elimination or intimidation of its enemies. Some para-institutional groups may have legal status as private, state-chartered organizations that are nevertheless led, organized, and manned by agents of the state itself. Others operate without any such charter, even though they typically operate on behalf of some or all of the state's coercive agencies and under their informal (if partial) sanction."[30]

This definition is broad enough to encompass both direct and indirect connections between the US and US-supported para-institutional actors. This might include mobilization, financial support, training, collaboration, turning a blind eye to coercive activities, or some combination of these different levels of collusion. Yet it is sufficiently focused to capture a variety of non-state military actors and the ways in which they can support, facilitate, or serve as surrogates or proxy agents. While para-institutional actors may be officially sanctioned groups such as PMCs, they can also include loose alliances and acquiescent arrangements with (possibly illegal) non-state armed militias or other similar forces.

The US military itself often refers to "irregular" forces. One US military document delineated that "irregular forces, are individuals or groups of individuals who are not members of a regular armed force, police, or other internal security force. They are usually non-state-sponsored and unconstrained by sovereign nation legalities and boundaries." Casting a wide net, such forces

> … may include, but are not limited to, specific paramilitary forces, contractors, individuals, businesses, foreign political organizations, resistance or insurgent organizations, expatriates, transnational terrorism adversaries, disillusioned transnational terrorism members, black marketers, and other social or political "undesirables."[31]

The Department of Defense (DoD) also often uses the term "para-military" to refer to "forces or groups distinct from the regular armed forces of any country, but resembling them in organization, equipment, training, or mission."[32]

The aim of this analysis of para-institutional actors in US foreign policy is to follow in the footsteps of recent scholarship that has pointed out the artificial dichotomous distinctions between "state" and "non-state," or the "public" and "private" realms, and instead highlighted the intersection of diverse militarized forces in the organization of power, particularly as it relates to the advancement of capitalism.[33] The distinctions between "state" and "non-state" and "public" and "private" force have often been blurred, as power is exercised through para-extensional means at both the global and local levels. US informal imperialism has been advanced via the orchestration or a bringing together, a collective synergy, of heterogeneous forces towards the stabilization of favorable political and economic environments in peripheral states. As we shall see, this coordination is largely performed through US state institutions, via military training and assistance, CIA and Special Forces liaisons, and the empowerment of some actors over others, but the broader picture is one in which the exercise of power transcends clear boundaries between state/non-state and public/private. Various forces assemble in the contestation of political and economic orders, processes of capital accumulation, and in the advancement and expansion of global capital.

What I refer to as *para-institutional complexes* or a *military-paramilitary nexus* encapsulates the multiple and myriad connections and alliances between the US state and a collection of para-institutional forces and the ways in which they intersect and come together in US imperialism. This goes beyond a "top-down" model or principal-agent conceptualization of the delegation of force in which the US creates or directly assigns activities to para-institutional agents. Para-institutional complexes includes a broader intersection of coercive forces in contexts of US statecraft. Rather than an "outsourcing" or "privatization" of US military force, such a nexus involves harnessing global and local agency, and a confluence of coercive actors orchestrated towards stabilization goals. Other scholars refer to "assemblages" or "webs" of security, police, and military forces to describe similar manifestations in the exercise of power in which an ad hoc and a contingent synergy between global and local forces form.[34] However, while these studies understand these assemblages as devoid of a central organizing power and detached from any specific programs, instead describing "organic" articulations of power, this book examines the connections between US imperialism and various para-institutional forces. It borrows from the idea of an "assemblage" to analyze the composite and flexible relationships that are increasingly built into US coercive statecraft and the processes of order-making that accompany it.

Layout and Overview of the Book

Chapter 1 outlines the drivers and contours of US imperialism. This provides a framework or lens through which to understand and explain US para-institutional complexes. It situates US foreign policy in relation to global capitalism and the ways in which the US state has served as a conduit through which the exigencies and interests of global capital and transnational elite classes have been advanced. Once these foundations are laid down, the chapter moves on to introduce the role of para-institutional forces in counterinsurgency and unconventional warfare.

The core of the historical analysis is presented in chapters 2–6. These chapters provide an historical overview of the development of para-institutional complexes in US statecraft. Chapter 2 concentrates on unconventional warfare and policies of regime change during the early phases of the Cold War. It argues that the desire for covert and deniable methods to overthrow unwanted regimes in the context of superpower rivalry, the strengthening international law, and the desire to preserve a US image of non-aggression gave rise to proxy forms of intervention. Chapter 3 focuses on US counterinsurgency training for allied states and US direct and indirect collaboration with localized militias, paramilitaries, and other similar actors. Chapter 4 examines the advancement of these practices during the Reagan administration. This provides the foundations for examination of the further entrenchment of these flexible sets of relationships in the immediate post-Cold War period in Chapter 5. Chapter 6 address the global proliferation of these networks during the US "War on Terror" and beyond.

Collectively, these core chapters present an historical account of the development of US imperial para-institutional complexes. They argue that para-institutional forces have become increasingly prevalent in US counterinsurgency and unconventional warfare from the covert Cold War period, to the entrenchment of these practices in the post-Cold War environment and their global proliferation in the "War on Terror." They show how these practices of leveraging of such forces have become increasingly institutionalized as part of standard doctrine and military policy, rather than merely a facet of covert warfare. They also demonstrate that certain historical events and trends have contributed into the overall receptivity of US planners over time to rely on para-institutional means to project US power abroad.

These chapters also identify US policy rationales behind the employment of PMCs, militias and paramilitary forces. An examination of US military doctrinal tactical rationales sheds light on the continuities and changes in US policies. In addition, threaded throughout each of these chapters is an overview of the continuity of US imperialism as well as its development in the context of globalization or an increasingly global capitalist system. While emphasis is placed on the continuation of US statecraft, and the role of para-institutional forces within these practices, these core chapters capture the evolution of the global context, particularly with respect to advances in the US maintenance role in the international system. In particular, they attempt to demonstrate how advancements in and the increasing transnational and global nature of capital have helped to entrench US outsourced imperialism. One way in which this has occurred, for example, hinges on the extent to which MNCs directly participate in stabilization interventions.

An acknowledgment is woven within these core chapters of how localized contexts and agency in various countries in the global South, such as colonial legacies, socio-economic inequalities and class hierarchies, amongst other factors have shaped para-institutional relations between US counterinsurgent campaigns and local collaborators. The para-institutional complexes described and analyzed throughout this book are not solely the product of a "top-down" application of counter-insurgent doctrine, but emerge within and are shaped by local contexts. These chapters acknowledge the interplay between the broader global structural pressures that have driven US imperialism and the local contingent nature of the manifestation of para-institutional complexes. The conclusion draws this analysis together to summarize some of the book's most significant observations.

US Imperial Statecraft and Para-Institutional Forces

What is US imperialism? What has informed it and what forms has it taken? Drawing from existing work on US imperialism, this chapter argues that throughout the post-World War II period US foreign policy has been geared towards expanding and preserving a US-led liberalized global order with the United States at its apex. In theorizing and describing the contours of an informal imperial project, this chapter provides an overview of US foreign policy that helps to explain US interventions into the global South. It highlights the contexts and structural driving logics that underpin US counterinsurgency and unconventional warfare programs. This sets up the argument that is laid out in the rest of the book: that para-institutional forces have become central features of these politico-military tools of statecraft and to US imperial strategies. This chapter, then, establishes how the development of outsourcing and alliances with para-institutional armed forces in US foreign policy in the post-WWII era is rooted in US imperial relations. This argument is then further developed throughout the subsequent historical chapters.

It is difficult to capture the multiplicity of factors at play in US foreign policy. There are many drivers of US policy, such as power politics (realist understandings of foreign policy and international relations), ideological sources (liberal perspectives), and material imperatives (Marxist-informed analyses), amongst others. There are many nuanced accounts of US foreign policy and these correspond to a huge variety of independent, and potentially intersecting, dynamics. Moreover, different international environments across time, such as the Cold War versus the "War on Terror," may provide very different regional and local contexts in which the US acts. Similarly, the enormity and complexity of US foreign policy in relation to the historical development of capitalism is difficult, if not impossible, to capture in one chapter alone. Providing this type of meta-explanation of US foreign policy runs the risk of an over-simplification of US foreign policy. The aim here, rather than detail a comprehensive theory of the sources and the nature of US foreign policy, is to provide a wider account of a US imperial project and how para-institutional forces function within it.

US Imperialism, Capitalism, and Global Order

Since its ascendancy to superpower status at the end of World War II, Washington has played the leading role in the creation and maintenance of an international framework in which the US and other states can pursue mutually reinforcing security and economic interests. A substantial part of this strategy has been geared towards the spread and defense of a global capitalist political economic system and support for corresponding forms of liberalized political and economic state organization abroad. The US has used its dominant position in the international system and its superior military capabilities to advance and preserve a liberalized global order as a whole within which it can pursue its interests but that is also beneficial to those interests of other core states, transnational class formations, and MNCs. These states, in turn, have increasingly looked to US leadership, as the most powerful state in the international system, to guarantee this system. The concretization of this into policy in Washington, or its strategic manifestation, is often referred to as the "Open Door" policy. The "Open Door" policy denotes an informal imperialism in which *de jure* sovereignty of other states is maintained (i.e. non-colonialism) alongside the promotion of "open" borders to free trade and access to local markets and resources.[1] The end "product" of global affairs this has produced (or "process" is better, as it is never complete but rather in continuous flux and growth and with multiple sites of "resistance") can be referred to as a US "empire" or a "US-led liberalized global order".

US imperialism is a consequence and part and parcel of the US state's relationship to the spread and defense of global capitalism. Many theorists have argued that the US has held a distinctive structural position in relation to capitalist expansion and processes of globalization.[2] Panitch and Gindin, for example, wrote that "The central place the United States now occupies within global capitalism rests on a particular convergence of structure and history," but more specifically this "was not a matter of teleology but of capitalist history."[3] With the rise of US power internationally in the post-WWII period, the US took over from previous European colonial powers as the leading capitalist state, the dominant military actor, and the leader of the "free world". Washington devised an international framework and rule-based system (following from Bretton Woods) that formed the basis of a US-led liberalized global order that replaced the Euro-centric system that preceded it.[4] Key to this has been a commitment in US foreign policy to the promotion of free trade and "open" borders encapsulated in articulations of an Open Door strategy. Scholars of US foreign policy have detailed the central managerial role the US has played as the lead architect and enforcer of this US-led liberalized global order. Michael Cox, for example, highlighted that the "underlying aim" in US post-war foreign policy has been to "create an environment in which democratic capitalism can flourish in a world in which the US still remains the dominant actor."[5] Andrew Bacevich agreed that US foreign policy is constituted by

> ... a commitment to global openness—removing barriers that inhibit the movement of goods, capital, ideas, and people. Its ultimate objective is the creation of an open and integrated international order based on the principles of democratic capitalism, with the United States as the ultimate guarantor of order and enforcer of norms.[6]

Similarly, Panitch and Gindin have written extensively about how the US has committed itself to "managing the international capitalist order" whereupon the US ensured and promoted "free trade and free enterprise internationally."[7] Simon Bromley has also argued elsewhere that the structures and logics driving much of US foreign policy can be observed within US strategic commitment to the defense and expansion of the wider global capitalist order and its dominance within that system.[8]

US policymakers in Washington have often reiterated the importance of this component of US grand strategy and the vision for a US-led liberalized world order. For example, the NSC-68 (1950), a document that essentially provided the blueprint for US Cold War foreign policy of containment, described this line of thinking: "we must make ourselves strong ... In the development of our military and economic strength," and that "Our overall policy at the present time may be described as one designed to foster a world environment in which the American system can survive and flourish," concluding that "we would probably pursue [this policy] even if there were no Soviet threat."[9] This has continued as one of Washington's official declaratory grand strategic policy aims beyond the Cold War. The 1997 *National Security Strategy for a New*

Century, for example, stated that at the core of the US "national security strategy" was a commitment to "promoting a world of open societies and open markets that is supportive of U.S. interests and consistent with American values." Moreover, this strategy depended, the report continued, on whether or not the United States is "able to sustain our military forces, foreign initiatives and global influence. It is that engagement and influence that helps ensure the world remains stable so that the international economic system can flourish."[10] US security strategy under Obama was not altered significantly in this respect, and his 2010 *National Security Strategy* detailed how the ultimate objective in US security strategy was to "underpin and sustain an international economic system that is critical to both our prosperity and to the peace and security of the world," by building international cooperation towards these goals.[11] This vision of the US's role in the world has been a fundamental and continual component of US imperialism and its corresponding Open Door grand strategy.

As part of this, US policymakers deemed that unencumbered access to resources, markets, and the labor of the South (or the Third World) was essential to both national security and the growth of the global economy.[12] US imperialism has been driven by long-standing geopolitical and economic imperatives in capitalist expansionism and accumulation to secure unfettered access to the raw materials and minerals, markets, and labor in the lubrication of the emerging global economy. The expansion of the US economy and the search of new markets for trade opportunities, investment and surplus absorption as well as global divisions of labor have long constituted a national basis for an imperial foreign policy. During the early stages of the Cold War, US planners also understood that "open" political economies in the South were essential for European post-war construction. The Marshall Plan and the economic integration of Germany and France devised to end inter-capitalist rivalry was also contingent on the continued flow of vital resources such as oil onto international markets. This perspective was a guiding rationale in one of the first approved national strategic outlines for US foreign policy towards the global South or Third World. President Kennedy's 1962 National Security Memorandum 182, titled *The United States Overseas Internal Defense Policy* (OIDP), made it clear that the US had a "political and ideological interest in assuring that developing nations evolve in a way that affords a congenial world environment for international cooperation and the growth of free institutions." This was partly because, it stated, Washington had an "economic interest in assuring that the resources and markets of the less developed world remain available to us and to other Free World countries."[13]

Contrary to classic notions of imperialism, such as those articulated by Lenin, US imperialism has not been driven solely by peripheral conquest for nationally based capital interests.[14] The US does not work to maintain a system solely beneficial to US-based corporations and US capitalist class interests against those of potential rival states (or foreign corporations or transnational elite interests), in much the way its colonial European predecessors tended to do. Consequently, the US has not locked itself into inter-imperialistic rivalry with other core capitalist states seeking to expand their respective national bases. Instead, as Doug Stokes has argued, the US has operated on a "dual logic" principle in which both national and transnational drivers underpin US Open Door policies. The United States

> … has long occupied a dual role that has been subject to both a "national" logic seeking to maximize US national interests and a "transnational" logic whereby it has played a coordinating role that has sought to reproduce a global political economy conducive to other core capitalist states.[15]

The US has worked to favor US-based capital interests as well as to maintain the functioning of an Open Door global order for the benefit of other core capitalist states and a transnational elite. In this way, for example, the US-led invasion of Iraq in 2003 was not about "blood for oil", as was popularly represented at the time by those critical of the intervention, subsequent occupation, and corporatist nation-building exercises. Instead, the US invasion and restructuring of the Iraqi political economy was to "open up" the Iraqi economy to global capital more generally and to ensure the stable flow of oil onto global markets (which, in turn, helped to stabilize oil prices around the world, and secured a diversified sourcing of global oil production, etc.).[16]

This "dual logic" to US imperial policies is also reflected in US policymakers' understandings of the role of US power. US policymakers understood that access to strategic national resources, such as oil, were of significant importance not only to the functioning and growth of the US economy but also to the advancement of the global economy. In a series of reports on the subject, US policymakers outlined the importance of securing flows of oil onto international markets for the US economy as well as those of other states. One report to the National Security Council in 1953 was acutely aware that

> Oil is vital to the United States and the rest of the free world both in peace and war. The complex industrial economies of the western world are absolutely dependent upon a continuing abundance of this essential source of energy … Since Venezuela and the Middle East are the only sources from which the free world's import requirements for petroleum can be supplied, these sources are necessary to continue the present economic and military efforts of the free world. It therefore follows that nothing can be allowed to interfere substantially with the availability of oil from those sources to the free world.[17]

This document continued to detail ways to secure the availability of oil for US and other core states' consumption. Another National Security Council report on US strategy towards the Middle East from 1958 outlined that "The Near East is of great strategic, political, and economic importance to the Free World. The area contains the greatest petroleum resources in the world and essential facilities for the transit of military forces and Free World commerce." The report further outlined that "The strategic resources are of such importance to the Free World, particularly Western Europe, that it is in the security interest of the United States to make every effort to insure that these resources will be available and will be used for strengthening the Free World."[18]

This US-led liberalized global order has evolved throughout the post-WWII era. Many theorists of global capitalism posit that the increasingly *transnational* nature of contemporary capitalism has gradually transformed a world economy into a truly global one within which the US state and international organizations (such as the World Bank, the WTO, and others) function as a stabilizer for an emerging transnational class.[19] For William I. Robinson, with the development of a global capitalism, particularly after the Cold War and with the rise of neoliberalism, the US state has increasingly occupied a distinctive position in the maintenance of global capitalism that has operated on behalf of transnational capitalist interests (rather than just those of US-based capital and elites). Investors from around the world, in the form of a transnational capitalist class, need to guarantee returns on their investments within and across various countries. A transnational authority is needed to ensure the protection of assets, capital, and property rights, and to safeguard sufficient stability for long-term projects. In part, this authority is constituted by international institutions the US helped to establish following the Bretton Woods agreements, such the International Monetary Fund (IMF) and the World Bank. However, it also has meant the need for a transnational coercive guarantee of last resort. In Robinson's words, "Global capitalism requires an apparatus of direct coercion to open up zones that may fall under renegade control, to impose order, and to repress rebellion when it threatens the stability or security of the system." Robinson continued to describe the US's role as "a *point of condensation* for dominant groups to resolve problems of global capitalism and for pressures to secure the legitimacy of the system overall."[20] The US state, and its military apparatus in particular, has embodied a congealment of transnational interests in global processes of capital expansion and accumulation in which the US, as the leading capitalist state and dominant military power, has served as a guarantor for the global capitalist system as a whole. The US has served as an instrument for transnational elites in managing the global capitalist system.

Crucially, this understanding of US imperialism presents a succinct explanation for broader continuities in US foreign policy throughout the post-WWII era. In rejecting fundamental ruptures in US foreign policy between distinct historical periods such as the Cold War and post-Cold War 1990s, and the "War on Terror", this analysis provides an account for continuities in US foreign policy, and those towards the global South in particular. It supports a continuity thesis that is based on a revisionist understanding of the Cold War, which in contrast to orthodox interpretations, posits that US policy towards the South was not exclusively shaped by containing an antagonistic Soviet threat and presence.[21] US policies of containment during the Cold War were not exclusively designed to contain the expansion of the Soviet threat. Instead, the East-West security tensions of the Cold War were largely subsidiary to the dynamics of North-South relations in which US policy was shaped predominantly by the logic of capital expansion and the need for the maintenance of "open" political economies in the South. As Layne explains, "Washington's ambitions were not driven by the Cold War but transcended it. The Cold War was superimposed on an existing hegemonic grand strategy that the United States would have pursued—or attempted to pursue— even if there had been no rivalry with the Soviet Union."[22] It is not that US-Soviet tensions did not play a role in US interventions in the global

South. They of course did. But rather that this picture is incomplete, and that salient dynamics in US imperialism and corresponding Open Door grand strategy provided the primary basis for Cold War interventions into the global South.

The US's Open Door strategies and its managerial role in the international system in "opening" political economies in the global South have continued after the end of the Cold War. The absence of a check and balance to US power in the international arena with the crumbling of the Soviet Union freed up the US to act on behalf of the accelerated pressures of globalization after 1991. The demise of the USSR removed the constraints of a bipolar world and allowed the US to continue to consolidate its position as the lead enforcer of the increasingly global capitalist system. The onset of the "War on Terror" has not fundamentally altered these deep-rooted sets of drivers of US foreign policy. For example, with the attacks of 9/11 heralding in a US re-armament and further militarization of its foreign policy, the ultimate and underlying goals of military expansion and increased intervention were not solely the destruction of Al-Qaeda and other terrorist organizations. Rather, the "War on Terror" was consistent with previous priorities in line with the US post-war grand strategy. The threat of terrorism has been used as a pretext (for which there is a long historical precedence)[23] to contain further threats to the stability of the functioning of the global order and to assert US primacy in the international arena. Therefore, in contrast to analyses that posit the "War on Terror" constitutes a new form of US imperialism, the "War on Terror" is another moment of US interventionism designed to advance the US-led liberalized global order.[24] This does not signify that terrorism and Al-Qaeda and ISIS do not represent a threat to US national security or that part of the "War on Terror" is not about counter-terrorism. Instead, it is within a specific context that these elements threaten the stability of a globalized liberal order. Ultimately, it is not the nature of the threat to the desired forms of "stability" that is important, whether this is communism, nationalism, or Islamic radicalism. Instead, it is the way in which such movements and agendas threaten the "stability" of desired parameters of the political and economic orientation of states in the South.

In addition, while the US has taken the lead in enforcing a liberalized global order on behalf of other states and the wider capitalist political economy, other core capitalist states have also played important roles in these processes of globalization. A wider constellation of core capitalist states, international institutions, and other agents have also contributed to the reinforcement of the global capitalist order and the advancement of global capital. States such as the United Kingdom (UK) and France, as two of the most prominent examples, have contributed towards the incorporation of peripheral or satellite states into this order.[25] Rather than a strictly hierarchical relation of international domination, US imperialism is embedded in broad and long-term political and economic processes both globally and in various locations of strategic interest around the world and has therefore also been dependent on divergent sets of agents (including allied states, political parties, militaries, class forces, etc.) for the advancing and the administration of this US-led liberalized global order.[26]

Refashioning and Stabilizing Countries in the Global South

Consistent with an Open Door notion of US imperialism, the promotion of liberalized political and economic state formations conducive to US interests and those of the wider global system has constituted the bedrock of US policy towards the global South. Policymakers in Washington have taken on a lead role in transforming and stabilizing the liberalization of political economies in the South towards US and transnational access to local resources and markets. As Panitch and Gindin put it, "The need to try to refashion all the states of the world so that they become at least minimally adequate for the administration of global order is now the central problem for the American state."[27] The US has operated to restructure and liberalize nascent state forms in the South in efforts to support their integration into the US-led global order which is conducive to the US's mutually reinforcing security and business interests and those of transnational capital. In this context, the United States' vision for its relations with the global South, as Kolko explained, "was far less the result of a conscious policy focused on the poorer and colonial regions than the by-product of its grand design for the entire global political and economic structure."[28]

Officials in Washington have implemented a variety of means to pursue its Open Door policies and to refashion states abroad for their incorporation into a US-led liberalized order. As David Painter has explained, "American officials deployed their nation's superior resources to ensure that the markets and raw materials of the periphery remained accessible to the industrial core of Western Europe and Japan as well as to the United States."[29] This has included various tactics and applications of power as well as diplomatic efforts, and debt and economic leverage, working through international financial institutions and other practices towards "opening up" states in the South as well as supporting the corresponding institutions needed to stabilize these arrangements abroad.[30] Some scholars have emphasized the US propagation of specific development paradigms and policies as part and parcel of American imperialism.[31] This included, for example, developmental models and the promotion of free trade agreements such as the Alliance for Progress towards Latin America during the Cold War as well as the promotion of a "Washington consensus" and neoliberal forms of governance in the 1990s. Tactics such as the use of structural adjustment policies (SAPs) and leveraging debt as well as the manipulation of international financial and monetary systems have also received substantial coverage as part of these processes.[32] In short, the US has worked to refashion states according to its Open Door vision for a US-led liberalized order through a variety of means.

However, a central challenge to these processes has often arisen from instabilities within countries in the global South. Challenges to the desired order have often taken on the form of social, political, and revolutionary calls for political and economic reform that threaten to steer countries along alternative developmental pathways and outside of the circuitry of global capital. In addition, violence and civil war have also often served to undermine stable business and market environments for long-term investment. Reformist and revolutionary movements, insurgencies, inimical reforms to political and economic policies in pro-US regimes, as well as general violence and lack of law and order have been of concern to US policymakers.[33] During the Cold War, this threat often took the form of communist revolutionary movements and armed insurgencies, but also included socialist democratic movements, resource nationalism, and other similar calls within countries in the South for more equitable distribution of wealth, national ownership of key industries, and protectionist tariffs and taxes, as well as a political and economic re-orientation away from the US-led liberalized order. Many countries in the global South during the Cold War were still European colonies, were undergoing a process of decolonization, or had recently gained their independence. In many of these places, years of colonial repression had produced highly unequal class configurations, poverty, brittle and unstable social bases, and internal conflict. People sought more than just independence from their former colonial masters, and Third World nationalism was often tied to larger anti-Western struggles. Either way, these developments, deemed detrimental to US and transnational capital interests, had to be contained.

In response, as demonstrated by Stokes and Raphael, "stability" has run through Washington's policy rhetoric towards the South, and in particular, but by no means limited to, towards those states rich in oil reserve deposits or significant material and geopolitical interests.[34] During the Cold War, this took the form of anti-communism embodied in the policy of containment. Containment often involved interventionist means of armoring "friendly" political-economic orientations, referred to as "foreign internal defense," or "stability operations". For instance, Kennedy's *Overseas Internal Defense Policy*, an official US foreign policy document outlining US policy stance towards internal instabilities in the global South, claimed that

> The susceptibility of developing societies to dissidence and violence which can be exploited by the communists requires the development of indigenous capabilities to cope with the threat to internal security in each of its forms. Reasonable stability is necessary for healthy economic growth, and the evolution of human liberties and representative government.[35]

An updated report authored during the Lyndon Johnson administration expressed a similar position:

> Internal security situations in certain developing countries are a matter of concern for the United States. Because of location of economic resources ... the United States must pay special attention to these countries and to the ability of their governments to maintain internal order. In certain circumstances, the United States may have to provide governments with assistance for internal defense purposes in order to help protect United States local and strategic interests which might be threatened by internal disorder or subversion.[36]

Concerns over the threat of "instabilities" in countries with vital and valuable resources were particularly prominent and highlight the centrality of policies of foreign internal defense in the US imperial project. Given the location of strategic resources, such as oil, rubber, certain minerals, and others, US policies were geared towards the stabilization of "open" political and economic arrangements and corresponding socio-economic relations conducive to transnational access to these resources and markets. A 1952 report titled *United States Objectives and Policies with Respect to the Arab States and Israel* elucidated some of these problems. This document outlined the geo-strategic importance of Middle Eastern oil, stating that

> ... more than a third of the world's known oil reserves are located in Arab states alone. Continued availability of oil from these sources ... are so important to the over-all position of the free world that it is in the security interest of the United States to take whatever appropriate measures it can, in light of its other commitments, to insure that these resources will be used for strengthening the free world.

However, the "nature of the problem", according to this document, came from internal "instabilities" in key countries in the Middle East posed by nationalist movements and inimical political developments. It stated, "the imminent threat to Western interests arises not so much from the threat of direct soviet military attack as from acute instability, anti-Western nationalism, and Arab-Israeli antagonism which could lead to disorder." A US response, primarily military in nature, "to prevent or overcome instability in the area," was therefore deemed "essential". The document outlined that given the impracticality of forging stability via "19th century fashion" military colonial conquest, partially because of new Cold War tensions with the Soviets, the US should provide military assistance and support to ruling classes for the "maintenance of stability in the area" and the "maximum assurance of independent regimes friendly to the West."[37]

In the post-Cold War environment, "stability" has continued to constitute a principal concern for policymakers in Washington. The 1997 *Quadrennial Defense Review*, for example, an official US military planning document, outlined priorities for "ensuring peace and stability in regions where the United States has vital or important interests and to broadening the community of free market democracies."[38] The *National Security Strategy* reports of the Bush and Obama administrations outlined US dependence on an open global economy and how "stability" abroad was essential to the US economy and security objectives.[39] US military reports in 2017 articulated similar positions. One 2017 military report highlighted that "a broad front of hostile challenges and forces are in position to sweep the status quo aside and in the process, create conditions that are profoundly unfavorable to U.S. interests." It categorized these as "revisionist, rejectionist, and revolutionary" forces (these included state and non-state challenges) throughout the world that impeded the US's ability to secure its strategic goals, one of which was to "Secure access to the global commons and strategic regions, markets, and resources." This document, moreover, reassured that "American military power does continue to insure or underwrite stability in critical regions of the world."[40] Other contemporary military documents have provided detailed instructions as to how to forge the desired stability towards US strategic interests.[41] Reference to this strategic ambition to maintain specific forms of "stability" in countries in the South will be threaded throughout this book in order to support these central claims about US policy towards the global South.

Efforts to forge and preserve specific forms of "stability" have often had to contend with long-term centrifugal social and political trends such as localized historical pressures, inequalities, and social dislocations which have helped to foment instabilities within the global South, including calls for political and economic reform considered inimical to US interests and the fluid flow of global capital. In refashioning states towards mutually supportive political and economic objectives, the US has had to contend with unfolding internal political dynamics within many countries that threatened to steer them away from the US sphere of influence and away from their integration into the US-led order. Undesirable political and economic changes, whether in the form of leftist armed insurgencies, popular socialist political and social movements, or other "radical" attempts to attain power, have presented significant obstacles to US Open Door strategies. The nationalization of key industries, closing local economies to foreign direct investment and absorption of surplus products from core capitalist countries, and the creation of exclusionary trading blocs (regionalism), and other similar moves have threatened to disrupt global capitalism. Westad summarizes this point:

> Third World domestic political conditions often needed to be changed first, before US-inspired reform could begin to take hold. Such change generally meant the defeat of radical attempts at controlling the political order, and it was in order to produce such a result that most US interventions took place.[42]

The deterrence of revolutionary movements and other nationalistic calls for change and the support for functioning liberalized political and economic architecture formed the basic frameworks of US "stability" operations.

The stabilization of favorable political and economic policies and frameworks abroad has often taken highly interventionist forms. US officials have often perceived it necessary to use coercive means to deter social and political forces within foreign countries that were perceived contrary to US objectives and those of global or transnational capital interests. Many scholars of US foreign policy further recognize the central role of US military power as a driver of global capitalism. Sullivan, for example, traced how US interventionism expanded in the post-1945 period throughout Europe, Asia and Latin America towards "opening up" economies in the South and stabilizing corresponding political and economic policies.[43] Such a conception of the role of the United States military was made explicit in Thomas Friedman's famous adage that "the hidden hand of the market will never work without a hidden fist ... And the hidden fist that keeps the world safe for Silicon Valley's technologies to flourish is called the U.S. Army, Air Force, Navy and Marine Corps."[44] Christopher Layne stated more simply that "US hard power forms the bedrock of an open economic system." "To preserve this needed geopolitical stability," he asserted, "the United States has taken on the role of hegemonic stabilizer in regions it has important economic interests ... to remove or block the coming to power of regimes whose policies are or would be inimical to openness, and to prop up friendly regimes."[45] Kolko also recognized the importance of US coercive statecraft in creating conditions congenial to the ultimate goal of global economic integration within countries in the South in which the US had substantial material or political interest.[46] Put differently, throughout the post-war era,

> ... the indigenous roots of low intensity conflicts throughout Third World societies produce an endemic security problem for the United States on local, regional, and global levels. As the architect, enforcer, and principal beneficiary of the new global order, unfavorable results at the low-level of conflict threaten United States access to strategic resources and undermine the global "suitable business environment" for profitable trade and investment.[47]

In the eyes of strategists in Washington, a militarized response to pacify growing discontents and deter moves to alter the developmental pathway of states away from US-led order has often been needed.

Counterinsurgency and Unconventional Warfare:
Coercive Strategies of Statecraft

Counterinsurgency and unconventional warfare have constituted the principal coercive modes of statecraft in the implementation of US Open Door grand strategy towards the global South.[48] US military assistance and training in counterinsurgency to allied regimes have historically constituted the primary politico-military means of stabilizing liberalized political economies abroad. US counterinsurgent military assistance to allied countries has helped to reorient recipient state militaries internally to oppose insurgent, political, and social groups deemed inimical to the prevailing order. The aim has been to stabilize "friendly" governments and insulate them from oppositional subversion and dissent "from below." Counterinsurgency constituted a form of coercive social engineering to simultaneously pacify dissent and forge consent for the prevailing order and its ancillary political and economic structures.

On the other hand, unconventional warfare has constituted US policies of regime change, but has also encompassed a variety of limited operations against unwanted foreign governments. While counterinsurgency has been aimed at stabilizing certain state forms and socio-economic relations employed as a response to internal "subversion" in pro-US or client states, unconventional warfare is the ultimate tactic in destabilizing "unfriendly" or "hostile" regimes. US planners have used it to weaken or depose leaders that direct (or threaten to direct) their countries towards alternate political and economic futures unfavorable to US interests and the demands of global capital circuitry. Successful unconventional warfare operations have the effect of ousting recalcitrant state leaders and replacing them with ones more malleable to US interests.

Both counterinsurgency military assistance to pro-US regimes as well as support to insurgent movements and other "irregular" forces against hostile regimes form a centerpiece among the coercive components of Washington's Open Door strategies towards the South. Counterinsurgency and unconventional warfare doctrine and training manuals have often indicated how these strategies have constituted politico-military coercive tools of statecraft designed to re-engineer states in the global South and bring them within US political and economic orbit. For example, the underlying rationales of "low intensity conflict" were articulated in a 1990 US military training manual, which included both counterinsurgency and unconventional warfare. It stated that losing "low intensity conflicts" would result in "unfavorable outcomes" including, among other things:

> The loss of US access to strategic energy reserves and other natural resources; The loss of US military basing, transit, and access rights; The movement of US friends and allies to positions of accommodation with hostile groups; The gain of long-term advantages for US adversaries.

Successful counterinsurgency or unconventional warfare, on the other hand, it stated, could "advance US international goals such as the growth of freedom, democratic institutions, and free market economies."[49] These arguments and examples will be developed in each of the subsequent chapters.

US policymakers understood that US military capabilities were not sufficient in themselves for the stabilization of peripheral countries and have worked to establish military-to-military ties with allied states for their own internal policing. President Eisenhower made it clear early on in the Cold War that the US "could not maintain old-fashioned forces all around the world," and that the US must "develop within the various areas and regions of the free world indigenous forces for the maintenance of order, the safeguarding of frontiers, and the provision of the bulk of ground capability."[50] Similarly, another US 1964 paper titled *U.S. Government Organization for Internal Defense* was clear that a central object of the United States should be

> To minimize the likelihood of direct U.S. military involvement in internal war by maximizing indigenous capabilities for identifying, preventing, and if necessary, defeating subversive insurgency, and by drawing on, as appropriate, the assistance of third countries and international organizations.[51]

Contemporary US "foreign internal defense" documents continue to provide a similar rationale for developing military-to-military ties to allied countries around the world.[52] One 2007 Air Force manual titled *Foreign Internal Defense*, for example, stated

> The US cannot unilaterally neutralize every significant terrorist threat or eliminate every guerrilla insurgency threatening the stability of friendly and allied nations. There are also serious legal and policy restrictions on what the US can do to directly counter or neutralize such non-state threats. Each host nation must eventually find its own winning combination of political, economic, informational, and military instruments to neutralize or eliminate internal threats. It is in direct US interest, however, to help these nations with various strategic, operational, and tactical initiatives.[53]

Consequently, the US has led the way in military training and financing towards the domestic or internal stabilization of order in countries around the world, but in those allied state formations in the South in particular.

Military assistance and training to strengthen a recipient state's capability for internal policing has been one of the most important features of US strategy towards allied states in the global South and has constituted a central mechanism through which US imperialism has traditionally been exerted. Planners in Washington have aided ruling classes and elite governing strata in "friendly" states to maintain "stability," with foreign military aid often directed towards "foreign internal defense." Between the end of World War II and the early 2000s, the US had spent over an estimated $300 billion in training and equipping around 2.3 million members of foreign militaries from all over the world.[54] The purpose of this training, according to Michael Parenti, has been "not to defend these nations from outside invasion but to protect ruling oligarchs and multinational corporate investors from the dangers of domestic anti-capitalist insurgency."[55] Courses in counterinsurgency and in unconventional warfare operations were taught in US military bases, in facilities across the globe, as well as on-the-location training programs. The School of the Americas (SOA—now renamed Western Hemisphere Institute for Security), for instance, was originally established at Fort Gulick in Panama in 1961 and then later moved to Fort Benning, Georgia; the school was responsible for training over 61,000 soldiers and civilians during the course of the Cold War.[56] Lora Lumpe found that "U.S. forces have been training approximately 100,000 foreign soldiers annually. This training takes place in at least 150 institutions within the U.S. and in 180 countries around the world." "Moreover," she asserted, "this training still focuses on central Cold War-era counterinsurgency doctrine—called foreign internal defense (FID)—rather than on new peacekeeping or defensive strategies."[57] According to another study, around 400,000 officers have graduated from US military institutions, the majority in counterinsurgency and related forms of stability operations between 1955 and 1981 alone.[58]

This is divided up into various training programs. Stokes and Raphael highlighted in 2010 that "the highest profile training program (International Military Education and Training, or IMET) has seen over 700,000 'friendly' officers pass through its courses since 1950, in an effort costing over $3 billion."[59] From 2001 to 2003, the IMET program grew 38 percent, from $58 million to $80 million, and expanded the delivery of training exercises from 96 countries in 1990, to 133 in 2002 as part of the "War on Terror."[60] This has grown significantly more recently, with the Government Accounting Office recording a rise in funding appropriated for IMET programs from around $62 million in 2000 to $108 million in 2010, and $109 million in 2015.[61] Although classes conducted as part of IMET have a broad range including language instruction and military resource management amongst many other possible courses, the purpose of IMET according to the State Department is to strengthen military-to-military ties with US allies and reinforce international security cooperation, but also to support the professional development of the recipient country's military for domestic stabilization.[62]

Another channel of military training, the Joint Combined Exchange Training program (JCET), was created in 1991 to allow Special Forces to hold joint training sessions in counterinsurgency and other related tactics with members of foreign militaries.[63] JCET, according to declassified US military reports, has increased significantly since the September 2001 attacks; it experienced significant growth in operations in 2008, and continued to accelerate under the Obama and Trump administrations.[64] Similarly, "Section 1206" of the National Defense Authorization Act for 2006 provided the Secretary of Defense a special mandate to authorize training programs in counterterrorism and stability operations. Through this additional channel of funneling military assistance, $1.574 billion was spent between FY 2006 and 2011 on training foreign forces primarily in the global South.[65] In 2013, the State Department recorded the US trained around 64,000 students from 152 countries, costing approximately $738.3 million for that year alone under the Foreign Military Training program.[66] These examples underscore the extent of US military assistance and how the fortification of pro-US governments' security forces through the provision of counterinsurgency assistance in order to police for internal insurgency, subversion, and unrest has been a significant facet of US foreign policy towards the South in the post-war era.

In accordance with continuities in US objectives and overall grand strategy, these coercive tools of statecraft used to forge the desired "stability" abroad have also remained relatively unchanged. Over time, the US military has ostensibly redefined its counterinsurgency and unconventional warfare and related military doctrines under various titles such as Small Wars, Counterinsurgency, Foreign Internal Defense (FID), Low Intensity Conflict (LIC), Military Operations Other Than War (MOOTW), Stability Operations, and most recently, in the "War on Terror," Irregular Warfare.[67] This shift in terminology, however, has not significantly altered US coercive strategies towards the South. On the contrary, while there have been nuanced adjustments to Washington's approach to counterinsurgency and unconventional warfare, in lieu of advancements in technology (such as the use of drones) and variations in the nature of the political threat (such as the rise of Islamic fundamentalism), amongst other considerations, there has been notable continuity in the core concepts underlying these forms of intervention, and the basic assumptions and principles that undergird the US approach, as well as in the methods they advocate.

There is also continuity in the way in which the US has used counterinsurgency and related forms of intervention in the South to respond to problems of "instability" and insurgency. This continuity of US counterinsurgency practices, in turn, has significant implications for the underlying prerogatives behind employing such forms of interventionism, as counterinsurgency specialist Hippler acknowledges:

> The conclusion we can deduct obviously is that the ideological settings colored and influenced military thinking and strategy, but that they were of little importance compared to the stable practice of US military intervention and the approaches in carrying them out. If US interventions and their strategies hardly change over time and happen independently of their ideological context, these ideological contexts cannot explain these interventions and their character.[68]

Similarly, Burnett and Whyte note an inconsistency between the identification of new types of warfare and the strategies implemented to address them in the "War on Terror":

> In political terms, the claim is that the new terrorism represents a break from the past ... [I]t is highly illuminating ... that we see, in the example of Iraq, a simultaneous call for a return to the old counterinsurgency strategy. Despite all of the hyperbole surrounding the new "netwar" and the new terrorism, it is being argued that this enemy should be dealt with in precisely the same manner as 20th century colonial rebellions.[69]

Para-Institutional Forces in Counterinsurgency and Unconventional Warfare

The US, and US-supported allied states, have increasingly forged composite relationships with various para-institutional forces in the prosecution of counterinsurgency and unconventional warfare. US imperial endeavors have enlisted the aid of para-institutional forces under various tactical and strategic rationales. These have also followed certain patterns of relationships. Private military companies (PMCs) have supplemented US capabilities and served as "force multipliers" at the global level or across different sites of US-led interventions as well as providing logistical, security, and combat support to local state and paramilitary forces. PMCs have allowed the US to maintain a military presence around the world and deploy US forces to multiple areas simultaneously by maintenance services for US troops, offering training, operating technical equipment and other tasks.[70] PMCs have often constituted a primary conduit through which US military assistance is provided, training and assisting foreign forces in their internal defense. Meanwhile, US-supported counterinsurgencies and unconventional warfare engagements have depended on local collaboration and militia forces. US practices of statecraft have helped shape para-institutional complexes of coercion—a nexus between state and non-state armed forces—towards the making and stabilization of particular types of political and economic order.

Practices of US unconventional warfare have involved complexes of para-institutional armies in which US military planners have mobilized, supported, and/or allowed insurgent forces to conduct violent acts to remove unwanted regimes from power. Such forces have included sponsoring of existing insurgent groups, inciting mass unrest, and the creation of mercenary armies to invade and overthrow hostile governments. US unconventional warfare doctrine over the last 60 or so years has advocated working "with and through" a collection of non-state armed forces such as guerrilla organizations, civilian armed militias, and private armies to overthrow "unfriendly" governments. US unconventional warfare doctrine has described in detail how US Special Forces and the CIA create and support a variety of armed forces towards destabilizing unwanted regimes in foreign countries.[71] US planners envisioned leveraging and assembling various insurgent assets simultaneously towards the orchestration of a military removal of an unwanted foreign government. For example, one Special Forces manual, *Army Special Operations Forces: Unconventional Warfare* (2008), outlined comprehensive instructions for the overthrow of foreign governments through supporting "irregular" forces and "surrogates" in support of US objectives. It cited prominent examples of successful US unconventional warfare:

> UW has been conducted in support of both an insurgency, such as the Contras in 1980s Nicaragua, and resistance movements to defeat an occupying power, such as the Mujahideen in 1980s Afghanistan ... more recently, SF operations in Operation ENDURING FREEDOM (OEF)/Afghanistan in 2001 and Operation IRAQI FREEDOM (OIF)/Iraq in 2003.[7]

US relationships with para-institutional forces in unconventional warfare have varied in terms of levels of US support. In some cases, US military and CIA operatives have directly mobilized, trained, and directed both global and local para-institutional partners towards the removal of "unfriendly" regimes. US forces have often sought to develop or manufacture an insurgency or uprising against unwanted foreign governments towards its eventual replacement with an incumbent government more pliant to US and transnational interests.[73] In other cases, US planners have supported pre-existing insurgencies. Special Forces and the CIA, among other agencies, have reinforced certain existing insurgent movements and guerrilla factions working towards objectives deemed beneficial to US interests.[74] Levels of support for insurgent or guerrilla factions have varied significantly from Special Force and CIA active participation, to the provision of substantial training, arms, and other materials to US-backed insurgents, to low levels of moral support and encouragement. The type and strength of the relationships formed has depended on variables such as the affinity between the US and local insurgents' political cause, the immediacy of the priority in overthrowing an unwanted regime, and the associated risks in providing support.[75] In conjunction with local forces, US planners often hired private companies to form the backbone of support to such "freedom fighters." PMCs have supported US unconventional warfare campaigns in a variety of capacities, such as logistics, training, and front-line combat. Chapter 2 of this book further details US support for guerrilla or insurgent forces and Chapters 4, 5, and 6 detail its development throughout the post-WWII era.

Similar arrangements have developed in US and US-supported counterinsurgencies. Counterinsurgency support and training to allied states advocated the deployment of pro-government militias, paramilitaries, mercenary groups, and other para-institutional forces to supplement US forces and the local incumbent government. US counterinsurgency doctrine has promoted the use of "irregular," "paramilitary," or "friendly guerrilla" forces and civilian "self-defense units" in support of "regular" armed forces.[76] This has included the promotion of paramilitary informant and intelligence programs for surveillance of suspect movements and populations. There is a general understanding reflected throughout US counterinsurgency training manuals and doctrines that para-institutional armed forces such as "militias, and other paramilitary organizations" can be effectively leveraged towards the counterinsurgency cause.[77] PMCs have provided an added layer to para-institutional complexes at various levels contributing to US and local state military operations in support roles. Crucially, PMCs have been increasingly hired by state forces and MNCs alike for security of important infrastructure, assets, and investments. PMCs have served as "investment enablers" in key regions, providing their clients with the protection they desire and helping to stabilize areas of economic importance. US and local states have presided over and coordinated concerted efforts between militias, mercenaries, and contractors towards stabilization goals with various complex relations between states, MNCs, PMCs and paramilitary forces. Chapter 3 provides further analysis of US counterinsurgency doctrine during the Cold War and the tactical logics that underpin paramilitarism contained in it. Chapters 4, 5, and 6 outline its development throughout the post-war era as an instrument of statecraft and the evolution in its use of various para-institutional formations.

There have been significant variations in the US and US-allied state alliances with para-institutional forces in counterinsurgent stabilization. In more direct connections, US Special Forces and the CIA have worked alongside "irregular" para-institutional groups towards US-supported counterinsurgencies. US training teams and advisors have supported local states and their militaries to create and control pro-government militia forces. In addition, US agencies have also formed indirect connections where the US relationship to localized para-institutional forces is mediated through the local state or military. US military planners train and support allied states and their official armed forces to leverage para-institutional groups, with limited or no direct connection between the US and the paramilitary agents operating on its behalf. In addition, US military assistance to a third state may also be channeled through PMCs, which are contracted either by US agencies or the local state military itself to provide training and other services that directly facilitate the local state's counterinsurgency initiative. Thus in some cases, while the US has had limited official relations with para-institutional forces, it nonetheless has enabled allied state structures to support their work towards beneficial results.

Militias, paramilitaries, and mercenaries and other factions in counterinsurgent settings have not always been products of counterinsurgent design. Instead, localized social and political historical processes provided the basis for the formation of these types of actors. Local contexts in countries in the global South, such as domestic class configurations, ethnic divides, social dislocations, and histories of civil conflict and civil wars, have also played a significant role in shaping the nature of US proxy engagements. Processes of capital accumulation as well as local elite formations, transnational class interests, and MNC pursuit of security for assets, land, infrastructure, or other interests play a role in the myriad of dynamics that help to shape local state relationships with paramilitary and militia-type actors towards stabilization.[78] These processes have unfolded in ways in which the production of "stability" and stable environments for capital growth has depended on multiple actors' participation in counterinsurgency and security. In other cases, long-term processes of state formation have contributed to the development of militia-type actors and vigilantes.[79] In addition, certain dynamics during conflict such as civilian security dilemmas and reactions to insurgent violence have also often precipitated in civilians forming village self-defense forces and other types of civilian-based militias in the context of contested political environments.[80] Either way, there are "bottom-up" processes of formation which color and influence the composite relationships with para-institutional complexes in US and US-backed statecraft.

In this case, in many of these para-extended proxy relationships, US and US-allied states might simply acquiesce or turn a blind eye to para-institutional actions taken on their behalf. It is often more appropriate to conceptualize the layered para-institutional force structures in US-led counterinsurgent stabilization as a military-paramilitary nexus, a series of connections between US, allied states, and paramilitary militia forces, rather than through a principal-agent paradigm in which US and allied states create and directly control a para-institutional coercive appendage to stabilization efforts. An examination of these relationships captures the broader nexus between the US and para-institutional actors and how they contribute towards the US-led imperial project.

These relationships are never fixed and undergo constant reformulation, as US governmental agencies and those of US-backed states amend or update their contracts with PMCs and alliances with paramilitary groups according to emerging circumstances. A military-paramilitary nexus model of US statecraft avoids an overly static understanding of power and the avenues through which it is applied. These para-institutional complexes have been comprised of multiple alliances, associations, and relationships between US, allied state forces, and para-institutional groups. This has functioned as a bespoke package of military relations emerging within and applied to local contexts in the conduct of US coercive statecraft. This analysis denotes a process of stabilization through para-institutional means rather than an off-the-shelf tool of military intervention.

In addition to these direct and indirect relations, the US has the power to help categorize which actors are conducive to global order and those that are inimical to it. US-led discursive constructions have aided in the de-legitimization of some para-institutional armed actors and in the legitimization of others. This has made it possible to target unwanted elements of foreign conflicts and turn a blind eye or acquiesce to others that have had objectives favorable to US Open Door strategies. For example, Reagan referred to fighting forces in Afghanistan, Nicaragua, and Angola as "freedom fighters," whilst helping to delegitimize other actors as "terrorists." Similar powerful discourses have been deployed in the "War on Terror" which have enabled the empowerment of some agents over others.[81] In this manner, US officials are able to, for example, simultaneously thank Iraqi militias for their services in support of US counterinsurgency and counterterrorism operations in northern Iraq and Syria, and deride Iran for supporting "terrorists" with similar objectives. Looking at the larger picture, the US has often been able to influence which actors and what actions are broadly perceived by international audiences to be legitimate (and those that are not) and thereby differentiate between those forces worthy of support and those the US and other powers can tolerate, and those worthy of condemnation and targeting.

Conclusion

US policymakers, as part of a broader Open Door grand strategy in US foreign policy, have worked to incorporate peripheral states, primarily in the global South, into a US-led liberalized global order throughout the post-war period. US imperial policies have been geared towards the rollback and containment of social and political forces that are deemed detrimental to US interests and those of global capital. As part of this, US policymakers have worked to stabilize the often tumultuous processes involved in the liberalization and transnationalization of political and economic frameworks in foreign countries. The fluid flow of natural resources, such as oil and other leading primary materials and commodities, onto international markets, and the creation of stable investment climates and the expansion of foreign markets have constituted fundamental objectives within this Open Door strategy. However, due to the unstable nature of brittle social bases in many peripheral states resulting from decolonization, massive inequalities in wealth, sharp class hierarchies, and exclusionary social systems, countries in the global South have been characterized by "instabilities," such as revolutions, social and political upheaval, and civil conflicts, as well as calls for political reforms deemed inimical to US and transnational interests. These instabilities have threatened the desired end-state of "open" or liberalized political economies and transnational access to resources and markets. In response, US policymakers have promoted coercive forms of intervention to maintain the desired forms of stability abroad. Counterinsurgency and unconventional warfare have represented the primary modes of coercive statecraft in forging and stabilizing the desired socio-economic relations and corresponding political and economic policies in areas where it has met resistance in the form of "radical" attempts at pushing for political and economic change.

As the subsequent chapters examine, US and US-supported statecraft have increasingly depended on collections of para-institutional forces. US statecraft has advocated the proxy use of global PMCs, mercenary outfits and local militias, guerrilla groups, paramilitary forces, and other para-institutional actors towards the suppression or elimination of counter-hegemonic forces unfavorable to US and transnational capital interests. Such forces have become integral to US imperial strategies and to forging or maintaining specific types of stability in countries in the global South. This has historically taken on two main forms. First, covert activity has been a mainstay of US policy towards much of the global South, with clandestine operators such as the CIA and Special Forces working with para-institutional armed forces towards shared objectives. Second, US military-to-military training in counterinsurgency to allied government military forces across the world has advocated the construction of paramilitary organizations to oppose internal or domestic dissent abroad. US and US-backed coercive practices to stabilize liberalized politico-economic orders abroad have been implemented via paramilitary extended means which have entailed the formation of significant alliances with a collection of para-institutional forces.

2

Covert Regime Change in the Early Cold War: "Power Moves Involved in the Overthrow of an Unfriendly Government"[1]

During the early Cold War, US military advisors conducted unconventional warfare to reverse unfavorable political and economic developments in foreign countries, typically in countries in the global South or the Third World. Unconventional warfare constituted the US method of choice to weaken or overthrow unwanted governments. It was designed to "roll back" governments deemed detrimental to US interests and those of global capital. Such strategies depended almost entirely on para-institutional complexes. US agencies liaised with and coordinated complexes of local collaborators, insurgents, militias, "secret" armies, mercenaries, private air-military contractors and other para-institutional forces to influence the political and economic orientation of foreign states. These efforts to weaken unwanted foreign governments or towards regime change extended around the world to countries such as Albania (1949–53), China (1949–60s), Burma (Myanmar) (1951–53), Tibet (1959–60s), Iran (1953), Guatemala (1954), Syria (1956–57), Egypt (1957), Indonesia (1957–58 and 1965), Iraq (1963), North Vietnam (1945–73), Cambodia (1955–70), Laos (1958–63), Cuba (1959–present), Chile (1964–73), Greece (1967), Bolivia (1971), Zaire (1975), Angola (1975, 1980s), Seychelles (1979–81), Libya (1980s), Grenada (1983), South Yemen (1982–84), Nicaragua (1981–90), (Afghanistan 1979–89), Fiji (1987), among others.[2]

These proxy forms of intervention developed within the context of covert action. Planners in Washington, in the desire to avoid Cold War superpower tensions and to conceal US interventions from international observers, devised a series of relationships in which the US could leverage the coercive capabilities of private military agents, insurgent forces, and local collaborators. The desire for plausible deniability as well as to distance the US from acts taken on its behalf channeled US policy through covert and outsourced means. US policymakers provided both the CIA (formed in 1948) and Special Forces (formed in 1952) with a specific mandate to undertake covert operations and to liaise with and support guerrilla factions. US covert paramilitary paradigms and unconventional warfare involved composite alliances with various para-institutional forces and became part and parcel of low-level conflict and US-directed subversion.

Specific patterns of relationships in US unconventional warfare developed, and aided the emergence of new types of actors to satisfy Washington's desires for deniable modes of statecraft. The CIA and Special Forces honed their skills at sponsoring local insurgents for subversion and the eventual overthrow of unwanted regimes, and in developing para-institutional armies to mount invasions of foreign states. Meanwhile, these covert operators sought out privatized methods of logistical and combat support. The CIA turned to airline businesses and retired military personnel and volunteers to form semi-private air-contractors which the US could hire for various covert and overt missions abroad. Semi-private air-contractors, such as Air America, Civil Air Transport (CAT) and Southern Air Transport, provided significant flexible, for-hire air force capabilities to buttress US unconventional warfare. The desire for covert and para-extended means of intervention led to the creation of private and semi-private for-hire military capabilities during the Cold War. These companies later formed the principal precursors to contemporary PMCs.

Unconventional warfare practices and US reliance on para-institutional complexes developed rapidly in the post-WWII period. By the early 1960s, entire wars, such as the Bay of Pigs invasion or the "secret war" in Laos, were orchestrated almost entirely via outsourced means by parallel networks of US-supported para-institutional agents. In many situations, the para-institutional forces supported were substantial in their size and scope. For instance, some estimates put the number of soldiers the CIA helped to mobilize and equip in Tibet to fight against the communist Chinese government at around 14,000 individuals.[3] Private airline companies contracting to the US, such as Air America, constituted some of the biggest airliners at the time with respect to how many aircraft it owned and operated.[4] As historian Greg Grandin put it, US counterinsurgency and unconventional warfare specialists in the 1950s and 1960s "transformed anti-communism from a parochial reflex into a world-historical paramilitary movement with pretensions no less universal than those of Marxism."[5]

US Imperialism and Unconventional Warfare

The advancement of a US-led liberalized global order was often challenged during the early Cold War by reformist or revolutionary foreign governments that implemented "radical" political and economic changes that brought their countries along alternative developmental pathways. A communist revolution in a given country, for example, represented a significant threat to MNC business interests as well as to the emerging global economy. Similarly, socialist governments and nationalist governments often instituted changes considered inimical to stable and open business environments. These often included state-led programs for the expropriation and nationalization of major industries, rural reforms, land redistribution, levying significant taxes on foreign companies and the redistribution of wealth, among other possible sets of policies which limited the "openness" of local markets as well as transnational access to raw materials, low-cost labor, free trade, and foreign investment. Moreover, of course, during the Cold War, these problems were compounded by security concerns and the potential for superpower confrontation when Soviet influence was added to this mix. In short, revolutionary and reformist governments often threatened to "close the door" on capital expansion and the vision of a US-led liberalized global order.

Washington devised a variety of responses to curtail and "roll back" such deleterious developments in the political economy of foreign countries. This was usually articulated and justified in the form of containing communism and supporting the "free world." The Truman Doctrine, for example, reflected orientations in US foreign policy towards the curtailment of communism and Soviet influence abroad and a commitment to assisting US-friendly governments. In cases where governments deemed unfavorable to US interests gained power, the go-to last resort was unconventional warfare to forcibly remove them.

Unconventional warfare amounted to coercive modes of statecraft and the engineering of politics abroad to replace governments which were uncooperative with those more amenable to US interests. The US unconventional warfare doctrine advocated supporting insurgency and guerrilla warfare for the eventual overthrow of foreign states.[6] One

CIA manual, *Power Moves Involved in the Overthrow of an Unfriendly Government*, provided detailed accounts of maneuvers to overthrow foreign unwanted regimes.[7] Special Force manuals contained similar instructions for accomplishing every phase of an insurgency.[8] US planners viewed unconventional warfare as the US equivalent or "mirror image" of communist popular revolution, perhaps best epitomized by Mao Zedong's *On Guerrilla Warfare* or Guevara's *Guerrilla Warfare*.[9] Tactics involved mobilizing the masses into action and called for "sabotage," "demolitions," selective assassinations, and disruptions, among many other violent methods of intimidating and destabilizing state enemies from within. Orchestrated primarily through the CIA and US Special Forces, US military planners enhanced their capabilities to coercively subvert regimes that put up significant obstacles or limitations to the expanding US-led global order.

US policymakers also understood the potential for relationships with underground resistance movements around the world to provide protection against a possible future takeover by "unfriendly" governments. US planners envisaged training "stay-behind" forces to extirpate an occupying power. A 1961–63 training document outlined that

> … Indigenous Special Force type units should be trained for operations within their own country as stay behind forces in seizing control of the government. These forces would be a nucleus upon which to develop a large-scale irregular force for the overthrow of the hostile regime or occupying government.

A similar plan revolved around maintaining contact with para-institutional forces in third countries in case an invasion using private armies was deemed necessary. This same document detailed that "unconventional warfare forces of a *similar ethnic grouping* could be trained in adjacent countries, within the United States, or in some other host country. These forces would be a deterrent to indirect aggression by hostile political forces."[10]

Some of the most prominent examples of the way US coercive statecraft functioned towards capital interest in this manner include US-supported regime change in Iran (1953), Guatemala (1954), Cuba (1959–61), and Chile (1973). In 1951, the Iranian Parliament voted to nationalize the country's oil industries and Iran elected Mohammed Mosaddeq as the country's new prime minister.[11] Mosaddeq promised to further strengthen democratic politics and to carry out a nationalization plan which would see Iran pay for the expropriation of foreign oil companies' assets (the Anglo-Persian Oil Company). Unable to broker a deal, the CIA and British intelligence services fomented a coup in 1953 which unseated Mosaddeq, re-installed the Shah (Iranian monarchy), and reversed the nationalization of oil.[12] Declassified CIA documentary history of these events revealed that US policymakers viewed the nationalization of the Iranian oil industry as a threat to global oil supplies as well as to stable oil prices.[13] Even though largely controlled and operated via British companies, US planners deemed it necessary to intervene on behalf of transnational interests in oil flows across the world.

A year later, 1954, in Guatemala, the US directed a coup that overthrew Guatemala's democratically elected leader Jacobo Arbenz. Numerous declassified official US government documents as well as a deluge of academic articles and books detail how the redistributive agrarian reform and nationalization efforts of the leftist Arbenz government threatened US business assets and interests.[14] To reverse these inimical reforms to Guatemalan economy and society, the US government ordered the CIA to orchestrate a coup against the Arbenz government. In Cuba, Castro's 1959 revolution deposed the right-wing authoritarian Batista regime and completely overhauled the economic model of the country with progressive social reforms, expropriation of land and rural reform, nationalization of key industries, and the expropriation of foreign owned assets. In response, the US mobilized a private army of Cuban elites and dissidents in attempts to invade Cuba and depose Castro's regime.[15]

MNCs played a central role in these processes. MNCs often pressured the CIA and the US government to take action against the promise or implementation of reforms to political and economic policies that might damage their interests. In Guatemala, United Fruit, a major banana-growing conglomerate, made use of their extensive connections in Washington to push for American intervention. It ultimately played an instrumental role in the overthrow of Arbenz by soliciting the US's support to protect the company's assets against land appropriation, as well as by helping to finance the Guatemalan generals who would go on to lead a US-led mercenary insurgency against the government.[16] In Iran, the Anglo-Persian Oil Company (which would later become British Petroleum) lobbied US and UK officials to take action against Mosaddeq's nationalization plans.[17]

In Chile, when the election of Salvador Allende running on a populist Socialist platform promised to nationalize major industries and provide platforms for redistribution of wealth, MNCs operating in the country protested, lobbying the US to take action and in some cases working with the CIA towards Allende's removal from power. US declassified documents demonstrated that US policymakers deemed Allende's Socialist policies, such as the nationalization of the copper industry, a threat to US and transnational business interests.[18] A Senate Select Committee report titled *Covert Action in Chile 1963–1973* detailed US and third-country interests as well as MNCs actions in Chile, stating that "Leaders of American multinational corporations with substantial interests in Chile, together with other American citizens concerned about what might happen to Chile in the event of an Allende victory, contacted U.S. government officials in order to make their views known."[19] Companies such as Pepsi, International Telephone and Telegraph, Inc. (ITT), and Anaconda Mining, as well as Chilean media outlets such as *El Mercurio*, among other corporations, liaised with the US and CIA to prevent Allende's election and to help his overthrow once he was elected.[20] The ITT in particular made donations to the CIA as well as to oppositional leaders towards these ends.[21] The US response was clear. In a secret meeting, President Nixon promised to take revenge and vowed to "make the economy scream" so as to apply pressure on Allende.[22] Meanwhile the CIA set about coordinating Allende's removal from power. According to CIA documents, the US made it clear that "It is firm and continuing policy that Allende be overthrown by a coup," adding that "It is imperative that these actions be implemented clandestinely and securely so that the USG [US government] and American hand be well hidden."[23]

Covert Operations

US leaders and advisors became quickly aware of the multiple Cold War dynamics that placed limitations on the US's ability to directly overthrow unwanted regimes or militarily influence the formation of favorable state arrangements abroad. Sending US troops abroad was often not an option. Cold War tensions with the USSR meant that US interventions to reverse communism abroad could escalate into a major confrontation, possibly a cataclysmic nuclear one, between the two superpowers. In the early phases of the Cold War, President Eisenhower reflected that, despite the need for intervention in the Third World, the United States "would not want to provoke all-out war." In the context of such tensions, US policymakers deemed it necessary to be prepared militarily. President Eisenhower reaffirmed, "we must avoid putting sizable forces into the area, with great logistical establishments behind them. We would be tied down and would be unable to meet rising problems in other areas."[24] US policymakers were also keen to preserve an image of non-intervention to international audiences and to avoid violating key international legal treaties the US had signed in the wake of World War II. Consequently, US policymakers deemed it critical for the role of the US to remain hidden, or at least to be able to divest responsibility if US involvement was not possible to conceal. Plausible deniability during the early Cold War became an operational necessity. Military planners had to devise alternative ways to assert influence abroad within the confines of the bipolar international order. In an adjustment to these Cold War tensions, US policymakers developed the more clandestine arts of covert operations, unconventional warfare, and asserting power "with or through" global and local para-institutional surrogates.[25] Working "with or through" a collection of global/local para-institutional forces made this possible, and actions to overthrow or undermine foreign governments were delegated to para-institutional forces, rather than undertaken by agents of the US state.

In the late 1940s and early 1950s, leaders in Washington began to develop the institutions to administer or implement outsourced strategies of statecraft. They built the administrative infrastructure for covert operations and the delegation of coercion to para-institutional forces.[26] For example, the creation of the CIA, through the National Security Act of 1947, involved a license for "covert action" and regime change to counter Soviet and communist activities whilst avoiding major war. A year later, the National Security Council Directive NSC10/2 (1948) established the paramilitary capacity of the CIA. The NSC10/2 detailed that CIA paramilitary operations consisted of

… preventive direct action, including sabotage, anti-sabotage, demolition and evacuation measures; subversion against hostile states, including assistance to underground resistance movements, guerrillas and refugee liberation groups, and support of indigenous anti-communist elements in threatened countries of the free world.

Crucially, this directive also specified that such actions were "in the interest of world peace and US national security," and should be "so planned and executed that any US Government responsibility for them is not evident to unauthorized persons and that if uncovered the US Government can plausibly disclaim any responsibility for them."[27] In 1947, George Kennan, the architect of US policies of containment, alongside members of the State Department and Defense Department jointly proposed the establishment of a "guerrilla warfare school" to combat communist political advances in foreign countries.[28] The programs taught at this academy were geared towards training in insurgency and related tactics. In 1952, Truman created the US Special Forces division with the specific mandate to support guerrilla forces abroad and conduct limited covert operations.

The unconventional warfare doctrine acknowledged that the dynamics of Cold War superpower tensions meant US military planners had to devise methods of covert action. One Special Forces training manual, for example, outlined how the direction of guerrilla forces was akin to a proxy intervention designed to distance the US from these acts and explicitly, to "avoid formal military confrontation."[29] Avoiding escalation of conflict with the USSR was calculated into the decisions to outsource the infamous (failed) Bay of Pigs invasion of Cuba in 1961. In the build-up to the invasion, Kennedy strictly prohibited the involvement of US troops explicitly to avoid major conflagration with the USSR and so as not to jeopardize US interests elsewhere.[30] In a report from presidential advisor Arthur Schlesinger (Jr.) to President Kennedy, Schlesinger claimed that secrecy was vital to any mission to depose Castro. If the US committed American troops to the invasion, he warned, it "would have presented the Soviet Union with an American Hungary" and a justification for Soviet reprisals. Therefore, Schlesinger argued, steps should be set in place to make it appear that the invading force of Cubans was operating under their own accord and not directed or supported by the US.[31]

US policymakers also wanted to avoid tarnishing the US's reputation for non-intervention and anti-colonialism and as a champion of human rights and "freedom." During the development of US covert capabilities in 1947, in a letter to then Secretary of Defense Forrestal, George Kennan related, in reference to Soviet and communist revolutionary warfare and insurgent tactics, that "I do not think the American people would ever approve of policies which rely fundamentally on similar methods." He continued: "I do feel, however, that there are cases where it might be essential to our security to fight fire with fire."[32] Covert action became a way to avoid domestic and international public scrutiny. The Kennedy administration's *Overseas Internal Defense Policy* stated that while executing US statecraft, "it is important for the U.S. to remain in the background ... [so as not to] expose the U.S. unnecessarily to charges of interventionism and colonialism."[33] The CIA was also clear that US Cold War warriors should elaborate a cover story for its unconventional warfare operations because "other friendly powers or Western countries ... would find serious objection if it became known that the United States covertly supported the overthrow of a small foreign government, despite its communist character."[34] The political costs of war and the preservation of an appearance of US non-intervention pressured US policymakers into engaging in covert action and the development of para-extended means through which to exert influence abroad. Operating through local forces distanced the US from interventionist charges.

The Bay of Pigs invasion, again, provides an apt example of these dynamics. In a memo to President Kennedy, senior aide Arthur Schlesinger underscored the extent to which policymakers in Washington valued the ability for the US to appear uninvolved to domestic and international audiences. It stated:

> Our problem is how to protect the ... impression of the United States as a mature and liberal nation, opposed to imperialism and colonialism and dedicated to justice, peace and freedom. The operational contribution to this effort—i.e. Cubanizing the operation and doing nothing which would be inconsistent with a spontaneous Cuban effort—has been worked out with skill and care.[35]

One of the primary concerns behind avoiding US direct complicity in the Bay of Pigs invasion and thus the delegation of the operation to para-institutional forces was preserving an image for other countries of US restraint. Colonel Hawkins, in his official military report on the lessons learnt from the failed invasion, goes as far as to argue that the preoccupation with maintaining a positive image obstructed operational success by limiting the number of US personnel and divesting responsibility to paramilitary fighters with less experience.[36]

Part of this concern was the lack of domestic and international support for conducting such an operation. It is worth citing the Schlesinger Memorandum at length here:

However "Cuban" the operation will seem to be, the US will be held accountable for it before the bar of world opinion: Our own press has seen to that. Beyond this, there is an obstinate fact: A great many people simply do not at this moment see that Cuba presents so grave and compelling a threat to our national security as to justify a course of action which much of the world will interpret as calculated aggression against a small nation in defiance both of treaty obligations and of the international standards we have repeatedly asserted against the communist world.[37]

Decisions to attempt to keep the US involvement secret and to delegate the entire operation to para-institutional forces for that purpose was partially based on the desire to maintain a positive image of the United States in the international arena. Making the operation covert by working through private entities was a strategic imperative. However, efforts to maintain the US's role as hidden were not always successful.[38] In the Cuban invasion, for example, plans for the invasion had leaked and were exposed through various media accounts of training locations in Guatemala and downed privatized reconnaissance planes.[39] In another infamous case, the US role in weakening the Sukarno regime in Indonesia was exposed by the downing of a private contractor (Allen Pope) in 1958, whose subsequent arrest by the Indonesian authorities precipitated an international crisis.[40] While the Indonesian officials had found detailed evidence, including contracts and flight logs of Pope's missions, the US sought to deny their involvement by declaring Pope was merely a "soldier of fortune," unaffiliated with the CIA and attracted to Indonesia solely by the prospect of making money.

The Agents of Imperialism

The implementation of unconventional warfare modes of statecraft involved the development of global/local para-institutional complexes for its execution. US unconventional warfare specialists envisioned working "with and through" different sets of actors including mobilizing the civil population into rebellion, liaising with guerrilla insurgents, hiring global/local mercenary forces, and contracting to PMCs. This led to the creation of specific types of para-institutional actors in service of US foreign policy objectives.

US unconventional warfare doctrine envisioned mobilizing civil populations into revolt against their government. According to CIA documents, policymakers envisioned the activation of a total resistance amongst the population in a "target" county by "stimulating the unorganized more or less apathetic majority of the people to political consciousness, then to passive resistance, and ultimately to aggressive action."[41] US doctrine contained detailed analysis as to different segments of the population that can be mobilized using psychological techniques, misinformation, and even coercion. Similarly, Special Forces training manuals identified civilian populations as "auxiliary" forces that would provide support to and participate in armed uprising.[42] Entire populations were viewed as a source of potential insurgent action. For example, the CIA distributed leaflets in Nicaragua in the 1980s to Contra forces and to civilians, that instructed citizens to take part in acts of opposition against the existing government including petrol-bombing police stations, amongst other acts of "sabotage."[43] Training manuals and instructional handbooks used to educate US and foreign personnel in unconventional tactics advocated mobilizing civilians and even schoolchildren as informants, logistical support mechanisms, and underground cadres of an overall resistance.[44]

The US doctrine made clear that the "most vital functions" in marshaling mass participation in a revolt is preserving "plausible denial."[45] Mass participation helped to make it appear as though domestic populations rose up and took action themselves as a result of the accumulation of discontent, rather than due to external or foreign tinkering in the politics of a sovereign nation. Another part of the rationale or logic in encouraging mass participation in the violent overthrow of a regime reflected a "mirror image" of communist guerrilla warfare and building popular support for particular political orders. Rather than merely militarily destroying or toppling a regime, US planners hoped to foment dissent and opposition to governments they deemed unfit, and eventually towards establishing popular support and a modicum of legitimacy for whatever government replaced the old. Local collaboration was considered necessary. For example, in the build-up to the 1953 overthrow of the Mosaddeq government in Iran, the State Department (1953) wrote that it "would be literally fatal to any non-communist successor to Mosaddeq if the Iranian public gained an impression that the new premier was a 'foreign tool.'"[46] After the coup, the Western-allied Shah re-took power and with the help of the US and British intelligence services, considerable efforts were made to make it appear as if the coup had resulted from a "spontaneous" and "domestic popular" revolution.[47]

These strategies also relied on oppositional forces such as armed political organizations, or thugs, or militarized labor unions, in a target state to conduct violent (as well as non-violent) activities that favored US interests. An official review, part of the Church Committee, of the CIA's activities in the build-up to the overthrow of Chile's democratically elected Allende in 1973 revealed that the CIA supported various "private sector groups," many of which were engaged in violent tactics.[48] Militant trade unions and other workers' organizations were directed or influenced into taking both non-violent and violent disruption. Moreover, the report noted that the CIA indirectly supported violent organizations. It stated that there were clear links between "CIA-supported political parties, the various militant trade associations (*gremios*), and paramilitary groups prone to terrorism."[49] In other words, the CIA worked through certain political groups, but indirectly allowed or acquiesced to wider networks of opposition to Allende's socialist government.

In a similar fashion, in order to obscure the role of the US in the 1953 coup in Iran against Mosaddeq, who two years prior had nationalized Iranian oil from British oil companies, CIA operatives mounted a coup using a variety of local collaborators.[50] CIA declassified documents, one titled *Campaign to Install Pro-Western Government in Iran* revealed how plans hinged on mobilizing small armed gangs and harnessing the defection of elements of the Iranian military.[51] Rahnema's book, *Behind the 1953 Coup in Iran: Thugs, Turncoats, Soldiers, and Spooks*, meticulously detailed the way in which the CIA (and British intelligence) orchestrated the overthrow of Mosaddeq, alongside local elite political forces who conspired to mobilize local agents. Rahnema noted how American clandestine agents helped in "mobilizing and coordinating the three networks of thugs, anti-Mosaddeq army officers and militant members of organizations and parties closely associated with prominent clerical, civilian and military anti-Mosaddeq leaders."[52] A loose coalition of para-institutional forces, including armed "ruffians," gangs, armed minor political organizations, and defected police and military were instrumental in engineering an uprising against Mosaddeq and re-installing the Shah to power, in line with the imperial interests of the US and the UK.

The principal coercive forces to spearhead US unconventional warfare operations and policies of regime change were US-supported guerrillas or insurgents. US military unconventional warfare training manuals during the Cold War (as well as beyond) made it clear that localized guerrilla armies should form the sharp end of US regime-change tactics. One Special Forces training manual detailed that the "primary mission of Special Forces units is to develop, organize, equip, train, support, and control guerrilla forces and to conduct guerrilla warfare." This document continued to outline how a "complete integration of U. S. Army personnel with indigenous guerrillas," to "work, fight, and live with the indigenous personnel" is needed to effectively overthrow an "unfriendly regime."[53] As such, training resistance forces and paramilitary action occupied a broad spectrum of activities including "political action, psychological operations, espionage, sabotage, assassination, traffic in contraband, and the gathering of intelligence."[54] According to other US military and CIA documents, local armed groups were organized into "saboteurs," "shock teams," "shock troops," and "tactical psy-war teams," and such groups were trained in the use of improvised explosives, booby traps, and other unorthodox munitions.[55] For example, one detailed how "sabotage teams" could be directed to a "range of suitable targets for sabotage … [including] telephone, telegraph, teletype facilities, electric power sources, or facilities, gasoline storage depots, military vehicles, railroad and highway bridges at selected sites, radio broadcast facilities and other facilities."[56]

Various instrumental rationales meant US policymakers prioritized local guerrillas over US official forces for unconventional warfare operations. Local armed groups had the knowledge and information of the language, terrain and of the people necessary to complete important tasks.[57] US doctrine outlined how local guerrilla forces are not only familiar with the local political situation, but can also help induce the defection of the populace against the incumbent government and create legitimacy for the post-coup order.[58] Also, US advisors often saw outsourcing to local guerrilla factions as a faster, more efficient and cheaper way to intervene compared to mobilizing US troops. As former CIA Director Richard Helms pointed out, in reference to the "secret" Hmong warriors in Laos, mobilizing and paying local armed cadres "was a much cheaper and better way to fight a war in Southeast Asia than to commit American troops."[59]

Similarly, "stay-behind" forces were positioned in "friendly" states in order to help ward off and protect incipient state forms against inimical government reform. One early example of these types of measures occurred not in US policy towards areas of the global South in response to Soviet machinations, but in post-WWII efforts to defend European countries against detrimental autochthonous political developments. President Truman's authorization for the CIA, in close collaboration with British MI6 and NATO, to help create, train, and arm paramilitary forces throughout much of Western Europe is instructive. Ostensibly designed as a dormant stay-behind insurgent force with hidden caches of weapons sprawled across many countries in Europe in case of a Soviet invasion, these paramilitary forces instead helped steer some European countries away from a communist political path when such a Soviet invasion failed to materialize.[60] According to Ganser, these civilian secret armies were

> ... involved in a whole series of terrorist operations and human rights violations that they wrongly blamed on the communists in order to discredit the left at the polls. The operations always aimed at spreading maximum fear among the population and ranged from bomb massacres in trains and market squares (Italy), the use of systematic torture of opponents of the regime (Turkey), the support for right-wing coup d'états (Greece and Turkey), to the smashing of opposition groups (Portugal and Spain).[61]

These forces remained intact, operational, and hidden up until 1990 when threads of such a paramilitary network began to be unraveled, with the last known meeting of the organizational committee in 1990 in Brussels. Very similar forces were secretly supported in the late 1940s and early '50s throughout much of the Baltic, some of Eastern Europe (such as partisan movements in Ukraine) and in Russia itself.[62] This type of operation followed the US unconventional warfare model in which "irregular" forces were commissioned to conduct military operations to influence political developments favorable to their US sponsors. They also set the stage for similar paramilitary actions throughout much of the global South in countries threatened by "communist subversion." However, it was not until the Eisenhower administration that such operations formed a principal component of the exercise of US power.[63]

Global and local mercenary outfits were another feature of US early Cold War covert unconventional warfare activities.[64] As Voß argued, the desire for use of ex-military expertise and former military personnel in a manner that would preserve plausible deniability was central to US deployment of mercenary groups throughout the Cold War. In many cases, US personnel were "sheep-dipped" or "sanitized," a process by which they were officially removed from their public duties so that they could perform deniable missions as "private" mercenary actors. Such forces often constituted a pivotal facet of US covert unconventional warfare policies, from Cuba, Congo, and Angola to Nicaragua, among many others, in support of the imperial project.

The preoccupation with keeping US interventions abroad covert also aided in the creation of novel militarized actors. In search of logistical capabilities for covert activities, US planners developed a network of semi-private military airline companies, including Civil Air Transport (CAT) and later Air America (AA) amongst many others.[65] These companies became the principal support infrastructure for unconventional warfare and the main point of contact with localized paramilitary forces, militias, and mercenaries, and often participated in combat operations. This ensured that entire operations were outsourced, with both logistical and combat initiatives delegated to non-state forces. These companies represent the principal precursors to contemporary PMCs and informed current US contracting practices. Building on the success of government contracting, many of these companies evolved into prominent PMCs, while others went on to become instrumental in the CIA's rendition program in the "War on Terror."[66]

Civil Air Transport (CAT), one of the first of these private air-contractors, began as the Flying Tigers, an "American volunteer group" composed of former US military personnel, fighting for the Chinese during World War II.[67] In 1950, according to recently declassified documents, the CIA and US Armed Forces capitalized on the covert logistical potential of airline businesses.[68] The CIA quickly expanded contracts, eventually bought shares in some companies (some became CIA front "proprietaries") and managed through the Pacific Corporation. Such airlines had various relationships with US agencies. As Robbins records: "Sometimes an airline would be wholly owned by the CIA, like AA, sometimes it would be partially funded by the Agency, and sometimes it could just be counted on for favors."[69] Air America (AA), the principal and most famous CIA "proprietary," was established by a former CIA agent, George Doole, as a private company.[70] A 1966 *Newsweek* article reported that "although in practical terms it is an operating arm of the CIA, AA is owned by a private aviation investment concern called Pacific Corporation."[71] Southern Air Transport (SAT) initially had an arrangement with the CIA that gave it ownership of only half its shares.[72] Finally, numerous other airlines were supported and contracted by the CIA and other US government agencies during the Cold War, even including Continental Airlines, Northwest, Air Ethiopia, and Air Jordan.[73] In short, the entire CIA air wing, or "proprietaries" as they were often referred to (somewhat erroneously), was essentially a vast collection of private airline companies with various ties to the Agency and the US government.

Similar to contemporary PMCs, these airline contractors constituted a privatized extension of US agencies. Air-contractors were often started, owned, and operated by retired military personnel.[74] This meant that rather than selling services to the highest bidder, CIA airline companies remained an instrument of US power. As Robbins describes, the pilots of CIA airline proprieties saw themselves as extensions to US policy rather than mercenaries, playing crucial roles in achieving US objectives.[75] This is similar to contemporary PMCs such as Military Professional Resources Incorporated (MPRI) which is noted for its "loyalty to US foreign policy objectives," and for employing a high number of US generals with the firm's headquarters located a few miles from the Pentagon.[76] In this way, these public-private partnerships served as a parallel network with policy direction firmly situated in Washington.

Such enterprises (as well as other PMCs) represented a significant component of US para-institutional complexes and were contracted for a variety of missions both covert and overt by various branches of the US government throughout the Cold War. A CIA training manual provided specific instructions for the use of such "private" airliners in US regime change. It outlined how logistical and air-fighter and bomber support to insurgent forces "should be accomplished in sterile aircraft only or by 'black flights.'" It also discussed how foreign pilots should be recruited and, for "maximum security," should be "unwitting of any official support of the operation."[77] Public-private partnerships of this kind were an integral component in US paramilitary operations, delivering supplies to, training, and sometimes leading combat missions, militarily supporting US-sponsored guerrilla armies and paramilitary assets in Burma (Chinese nationalists brought to the country to invade China) (1951–61), Guatemala (1954), Indonesia (1958), Laos (1955–74), Tibet (1956–66), Cuba (1961), Vietnam (1962–75), and many other locations.[78] The 1975 Church Committee (a Senate inquiry into intelligence activities of the previous two decades) concluded that these companies were important components of US foreign policy in general (rather than simply tools of CIA covert activity), stating that the "use and past expansion of the proprietaries was a direct result of demands placed upon the Agency by Presidents, Secretaries of State and the policy mechanisms of government. This is particularly true of the large air proprietary complex used to support paramilitary operations in Southeast Asia."[79]

Para-Institutional Complexes in US Unconventional Warfare

US regime change policies followed particular patterns and sets of public-private relationships in the early Cold War period. The pursuit of plausible denial meant the CIA and Special Forces worked in the background to assemble and coordinate complexes of para-institutional forces towards undermining or deposing unwanted regimes. Private logistical mechanisms and mercenary forces were deployed to support local resistance efforts. However, local contexts conditioned the mobilization of various non-state forces, their activities, and their relationships with one another. US imperial statecraft developed flexible para-institutional complexes organized around local conditions.

US action against the Arbenz government in Guatemala is a good example. Arbenz was elected president of Guatemala in 1951 elections, promising

> … to convert our country from a dependent nation with a semi-colonial economy into an economically independent country; to convert Guatemala from a country bound by a predominantly feudal economy into a modern capitalist state; and to make this transformation in a way that will raise the standard of living of the great mass of our people to the highest level.[80]

This hardly revolutionary agenda challenged the dominance of US corporations in the country, which owned and ran significant portions of Guatemala's industries. In particular, rural reforms and land redistribution efforts threatened United Fruit Company's land holdings. The US government, mainly through the CIA, set plans in motion (Operation PBSUCCESS) to overthrow Arbenz and install a successor more congenial to US economic interests.[81]

In order to influence the political direction of this small Central American country, the US mobilized various Guatemalan anti-communist fighters and foreign mercenaries from neighboring countries, and contracted a variety of airline companies to form a private air force.[82] The then head of the CIA Allen Dulles in conjunction with the DoD and State Department oversaw the mobilization of an insurgent force, which was designed to look like a domestic uprising. The CIA named Carlos Castillo Armas, (anti-reformist/anti-Arbenz) former solder in exile in Honduras, as "liberator" and leader of this pro-US invasion guerrilla force. Armas had attempted to take power in the past, but US "massive external assistance, including foreign troops" was decisive.[83] The CIA went about locating, training, and funding a competent guerrilla army composed of anti-revolutionary Guatemalan émigrés and mercenary soldiers from neighboring Central American countries, and set up training centers and bases in Honduras and Nicaragua. A part of the CIA training and support infamously included kill-lists and instructions on effective assassination techniques.[84] This included plans for the deployment of "special 'K' groups" and disguised mercenary soldiers to assassinate unarmed suspected communists.[85] The invasion force represented an outsourced network of collaborators that the CIA and Armas assembled between 1953 and the invasion in 1954. Alongside this paramilitary capability, the CIA contracted CAT to help form a "Liberation Air Force" to conduct bombing raids on Guatemala.[86] The CIA hired pilots, some of them Chinese nationalists and others retired US military personnel from CAT.[87] The CAT combat air support was an important part of the insurrectionary force against Arbenz, bombing and attacking Guatemala's armed forces, government buildings, and oil refineries to cripple the economy. It had a powerful psychological effect on Arbenz as an indication of the will and power of the guerrilla army. Under significant military pressure, Arbenz and his administration eventually resigned, paving the path for the installation of a US-supported regime. The new leadership enjoyed very little popular support and Guatemala experienced 40 years of dictatorship complete with paramilitary death squads, torture, disappearances, a regime supported by continued US counterinsurgency assistance against internal dissent.[88]

A similar model was applied in Indonesia during attempts to weaken the Indonesian Communist Party (PKI) and towards the overthrow of the left-leaning, non-aligned, and Nationalist President Sukarno in 1957–59. US and British intelligence services were wary of a growing Communist Party's influence on Sukarno's rule and its potential impact on investments in the country. Initially, US efforts rested on election manipulation and the use of propaganda to stem the tide of communism and to prevent left-leaning elements from coming to power. When this did not work, President Eisenhower authorized CIA director Dulles to conduct covert action to subvert and ultimately assassinate Sukarno, which came to be known as Operation HAIK, under the condition that the hand of the US remained hidden. To do this, the CIA assembled a complex of various para-institutional forces reminiscent of engagements in Guatemala. The CIA strengthened local insurgent factions (notably the PERMESTA) on various key islands of the Indonesian archipelago, such as Sumatra and Sulawesi, via training programs and the provision of arms and other materials. The CIA added a layer of international mercenary forces to help fortify these local efforts. Finally, it contracted air-military companies, such as CAT, for logistical support to local insurgents, and to conduct combat and bombing missions from bases and landing strips inside Indonesia as well as from the Philippines, Singapore and other neighboring states. The required infrastructure to undertake such a massive program and substantial US presence rendered its deniability all but implausible. When private company operatives were found out to be working for the US, as in the famous cases of Allen Lawrence Pope, whose CIA-contracted plane was shot down after conducing bombing raids, the US quietly withdrew its plans for the forced removal of Sukarno.[89] However, once Suharto, a general in Sukarno's military, took power in a *coup d'état* in 1967, the US backed the brutal counterinsurgent program Suharto set up to rid Indonesia of any internal dissent to the US-friendly order established, which culminated in the genocidal murder of significant numbers of communists, their sympathizers, union workers and activists, intellectuals, and land reform activists.

The failed 1961 Bay of Pigs invasion is also representative of the way parallel armed forces developed from CIA activity to covertly steer the political development of a small nation towards conditions more favorable to US objectives and interests. The Eisenhower administration supported the Batista government in Cuba throughout much of the 1950s. US policymakers at that time viewed "economic and political stability in Cuba as a means for promoting both U.S. commerce and security in the Caribbean world."[90] However, Batista's reign was characterized by repression and unstable inequalities in wealth. In January 1959, Fidel Castro's forces overthrew the Batista government promising a more egalitarian economy and society. Once in power, Castro proceeded to violently rid Cuba of Batista supporters in the institution of a new political and economic order. Under his direct control of the economy, Cuba aimed at expropriating around $1 billion of American properties as part of an agrarian reform initiative.[91] These modifications to Cuban economy and society angered US and Western businesses in Cuba and challenged US "Open Door" visions. The Eisenhower administration set in motion plans for the removal of Castro and the reversal of the revolution's reforms.

Designed specifically to ensure the perception of US non-involvement, the entire Bay of Pigs operation was delegated to para-institutional forces outside conventional US military channels. On August 18, 1960, President Eisenhower authorized the CIA to train, equip, and direct a collection of Cuban paramilitary fighters (called "Brigade 2506") and instructed the Pentagon to aid the CIA in training these fighters in guerrilla warfare in secret training camps in the US and Guatemala.[92] The private army consisted of Cuban exiles and anti-communist elites. The CIA also supported oppositional groups inside Cuba in the Escambray mountains which conducted numerous attacks against the Castro regime.[93] They also employed private airline companies for logistics and airborne attacks. According to *A Program of Covert Action Against the Castro Regime*, a declassified document of the 5412 Committee (the US committee in charge of devising paramilitary action against Castro's Cuba), the primary means through which the "replacement of the Castro regime" was to be accomplished was to "induce, support and so far as possible direct action, both inside and outside of Cuba, by selected groups of Cubans of a sort that they might be expected to and could undertake on their own initiative" and "in such a manner to avoid any appearance of U.S. intervention."[94]

The desire to maintain the perception of US non-involvement was one of the principal factors in delegating the Cuban intervention to para-institutional forces. Both Eisenhower and Kennedy made efforts to ensure that no Americans were allowed in combat roles to prevent revealing the US's role in the invasion. As such, Allen Dulles (the then director of the CIA) approved the purchase of shares in Southern Air Transport (SAT) amounting to $307,506 for the operation.[95] SAT and other "non-attributable" airplanes were leased or bought from similar CIA-affiliated companies such as Air Asia and Air America. Official CIA histories released in 2016 outlined how the CIA contracted CAT to supply pilots, trainers who could train Cuban pilots in bombing and aerial combat, and airplanes, such as B-26s and C-46s, for the invasion.[96]

As American contractors were forbidden to pilot planes (although this occurred in some cases anyway in violation of Kennedy's requests), the CIA hired Chinese and Latin American pilots and trained Cuban nationals to serve as the pilots for the invasion force.[97] The CIA contracted these airlines for the transportation of Cuban paramilitary fighters to and from a secret training base in Guatemala, to supply Cuban guerrillas in the Escambray mountains, and to conduct aerial attacks and bombing raids during the invasion.[98] The primary support mechanisms for this operation consisted of private entities with contractual obligations under the CIA and DoD.

When President Kennedy authorized the invasion in April 1961, approximately 1,500 irregular Cuban troops attempted to storm a number of beaches in Cuba reinforced with a privately run air support and fleet of boats, complete with US gunboats which remained in the background.[99] Although the Bay of Pigs invasion failed to achieve the desired objectives, and US involvement was ultimately exposed, it represented a step towards the formation of a para-institutional infrastructure upon which the US would later increasingly rely.[100] The CIA purchase of SAT shares further established a precedent and basis to conduct covert aviation operations through private means. SAT would immediately go on to expand operations in South East Asia and Latin America, winning a $3.7 million contract with the US Air Force in the mid-1960s to transport passengers and cargo in Asia.[101] Contracts for private airlines affiliated with the CIA rapidly expanded in the early and mid-1960s and, according to a *Time Magazine* article, at its peak was around double the size of commercial airline TWA, "employing nearly 20,000 people (as many as the CIA itself) and operating some 200 planes."[102] SAT also continued to operate in Latin America in various missions, such as in Nicaragua, as well as further afield in Angola, amongst others. But at least as important were the long-lasting public-private connections in and out of the government that were forged through such public-private alliances. For instance, George Doole, who was the CEO of AA, CAT, and others, was not only a former US Army pilot, but also a CIA officer, working with private operators to manage and direct these airline assets.[103]

Furthermore, some of the surviving members of the Cuban paramilitary invasion force went on to serve as hired guns for the CIA and other US agencies. Others continued to operate under their own initiatives. Many individuals affiliated with the Bay of Pigs invasion continued anti-Castro activities alone with US acquiescence, rather than direct support. Successor anti-Castro groups emerged such as Coordination of United Revolutionary Organizations (CORU), Alpha 66 and Omega 7. Luis Posada, for example, perhaps the most famous former CIA asset (1961–67) as part of the Brigade 2506 invasion force at the Bay of Pigs and a trained demolition expert, was behind the bombing of a Cuban Airline killing 73 people in 1976, with, according to declassified documents, full CIA advanced knowledge of the plan.[104] Posada was also involved in a number of other bombing plots and terrorist attacks against Castro and Cuban nationals.[105] One of these included planting bombs in Panama for which he was arrested and sentenced to eight years in prison.[106] Despite this, he was kept in cahoots with the CIA and was later hired by Col. Oliver North to aid the Iran-Contra resupply efforts in the late 1980s.[107] Another "freedom fighter," ex-CIA operative, and friend of Posada, Orlando Bosch headed CORU, linked not only to the bombing of a Cuban airplane but also terrorist activities in the Miami area.[108] According to a 1993 US Department of Justice report, many of these actors operated under the assumption that they had tacit or implicit support from the US government, based on the fact that they had received training and direction to commit these types of acts in the past.[109]

Further afield, the so-called "secret war" in Laos, a piece in the larger jigsaw of US statecraft in Vietnam (discussed in Chapter 3), followed similar patterns. US policymakers sought to stem Laotian communist forces as well as control the Ho Chi Minh Trail which the communist North Vietnamese were using as supply routes through Laos and parts of Cambodia. To do this, the US primarily operated through proxy para-institutional means. In the 1950s, the US contracted and allowed the French to hire US air-contractors such as CAT and AA for covert operations and resupply missions to French military detachments.[110] By the late 1950s, Special Forces had initiated paramilitary programs. As US and foreign involvement and pressure built in Laos, the 1962 Geneva Agreement was established to guarantee Laotian neutrality. Yet both the North Vietnamese and the US circumvented the treaty.[111] The US, instead of intervening directly, violated the spirit of the Geneva Agreement by increasing its support and engagement with various guerrilla forces, militias, and mercenaries. US engagement in Laos grew to become one of the largest covert CIA paramilitary ventures in the organization's history and constituted an outsourced total war.

CIA and Special Forces liaised with and supported various local guerrilla groups.[112] The Hmong people, led by Vang Pao, epitomized these relationships. CIA and Special Forces recruited and trained local mercenary armies organized into "Special Guerrilla Units" primarily to combat the North Vietnamese Army along supply routes into southern Vietnam. According to some estimates, the Hmong army had over 40,000 recruits at its height in the 1960s.[113] Many were paid for their services as "force multipliers."[114] The CIA also hired Thai mercenary commandos.[115] CIA-contracted airlines provided aerial support to these forces. The Hmong almost entirely depended on Air America and Bird and Sons, which air-dropped food and equipment, and served as a paramilitary transport air-wing for the "secret army," as well as US trainers and officials. These contractors were often put in the line of fire and conducted military missions. CIA-contracted planes reportedly flew clandestine bombing raids over Laos, dropping "hot soup," a concoction similar to napalm, on communist forces.[116] According to an official CIA history, in 1964 Air America was "bombing and strafing enemy positions both east and west of the Plain of Jars" in place of US military force, to preserve the image of neutrality towards Laos.[117] The extent of the devastation for the Hmong people was immense, prompting hundreds of thousands of deaths and refugees: "For the Hmong, America's 'war on the cheap' was a costly ordeal."[118]

Conclusion

Communist, socialist and "left-leaning" revolutionary and reformist states often implemented political and economic reforms that ran counter to a US Open Door vision for global order. US unconventional warfare tactics functioned towards reversing these trends through CIA and Special Forces-orchestrated *coup d'états* and ultimately served as a stepping stone towards the "opening" of these states' economies in order to bring them into the US sphere of influence. In order to hide US interventions abroad, US policies of regime change were delegated to collections of para-institutional actors. From US covert interventions in Latin America and South East Asia, the spread and defense of capitalist state arrangements abroad were spearheaded by parallel military and social forces. The delegation of force to para-institutional groups formed the backbone of many US covert operations offering the plausible deniability and avoiding the political complications that the commitment of

US troops would not. US outsourcing and delegation of coercive tasks developed within the context of covert action.

US outsourcing during this early period followed fairly consistent patterns and led to the development of a privatized CIA air-contractor multinational syndicate which formed the principal precursors to modern PMCs. The prosecution of unconventional warfare modes of statecraft usually depended on para-institutional complexes of local militias, insurgents, and guerrilla forces, as well as mercenaries and supporting contractors. US advisors worked directly to instigate and support agitators, armed protestors, and gangs, as well as mobilize, train, and equip insurgent armies. US agencies also employed mercenaries and hired contractors, mainly for covert logistical and support roles but also often in combat roles. These complexes of overlapping and coordinated flexible relationships were configured within local contexts. US agencies, and the CIA in particular, found contracting to companies an expedient and covert way to reinforce US paramilitary capabilities. These public-private ventures expanded quickly in the repertoire of US instruments of statecraft. Much like their PMC successors, they allowed US military personnel and expertise to be recycled through non-state institutions.

These interventions were often considerable in size and scope and all had significant implications for targeted countries. They played a central role in local contentious politics and in many cases determined the fate of entire countries. US interference (as well as pro-communist Soviet or Cuban influence, in some cases) in the form of unconventional warfare and support for coercive para-institutional force constituted socio-political engineering abroad at its best. These practices developed throughout the 1980s including in Nicaragua (1981–89), Afghanistan (1979–90), and Angola (1975–76).[119] And while Cold War warriors found increasingly effective and efficient ways to bring down unwanted regimes, US policymakers and advisors deployed similar methods of statecraft in the counterinsurgent containment of counter-hegemonic forces in allied states.

3

Counterinsurgent Statecraft: Militias, Mercenaries, and Contractors

In addition to overthrowing regimes "unfriendly" to US interests, the US has played the lead managerial role internationally in armoring processes of capitalist development and shielding transnational access to resources, economies, and labor. Pressure for reforms considered inimical to global capital, emerging from communism, socialist ideals, or nationalist political and social movements posed significant threats to "open" political economies in many countries in the global South or Third World. Given the threat of deleterious political and economic change "from below," the stabilization of pro-US regimes formed a centerpiece of US imperialism towards the South during the Cold War. This meant working with or through the ruling classes and elite political and economic strata in "friendly" states to insulate their governments and societies from inimical ideological influence. Foreign internal defense became a central plank of the broader strategy of containment. For instance, the *United States Overseas Internal Defense Policy* outlined how the US's role was to "assist in the immunization of vulnerable societies" and to "assist countries where subversive insurgency is latent or incipient to defeat the threat by removing its causes before the stage of insurgency is reached."[1] This took the form of bolstering foreign military capabilities for internal policing. US foreign policy was marked by the provision of substantial military assistance and training to allied state military forces.

Within the context of these programs, US agencies, mainly the Special Forces and CIA, often directly mobilized and supported para-institutional forces to complement local state-led or "host nation" counterinsurgent efforts. More indirectly, "friendly" governments developed and/or backed para-institutional forces under US instruction or recommendation. US military doctrines imparted to foreign countries through military training and funding emphasized support for militias, paramilitaries, civilian defense forces, armed thugs, and other "irregular" forces.

There is an extended history of a variety of non-state armed agents operating in concert with "official" forces in US-directed or US-supported counterinsurgency campaigns, including in Greece,[2] Italy,[3] the Philippines,[4] Thailand,[5] Korea,[6] Guatemala,[7] Indonesia,[8] El Salvador,[9] Colombia,[10] Chile,[11] throughout much of South America including in Operation Condor,[12] among many others.[13] Mercenary outfits and PMCs often supplemented these counterinsurgent efforts.

Often understood merely as a proxy war in containing Soviet influence, these extended para-institutional networks contributed to state formation and the consolidation of particular forms of order abroad. Much like in unconventional warfare, the origins of outsourcing and alliances with various para-institutional formations in US counterinsurgency were bound to US managerial roles in the international system and Open Door strategies. Para-institutional formations provided para-extensional means to participate in the processes of state development in countries in the Third World while US official armed forces "remained in the background." Para-institutional complexes connected to American-backed power centers were part of processes of order-making in the global South.

This chapter does three things. First, it examines US counterinsurgent doctrine and the instrumental rationales behind the use of para-institutional forces. It then moves on to briefly examine some examples of the application of this doctrine and the relationships that comprised US statecraft and the formation of para-institutional forces. It acknowledges localized agency and dynamics in these processes. The main argument is that the Vietnam War represented a turning point in the development of outsourcing, privatization, and US alliances with para-institutional armed groups. At the outset, US intervention into Vietnam (and Laos) followed similar patterns of covert activities and delegation of force. With increased US involvement and US troops on the ground in Vietnam (1965–73), the formation of public/private partnerships and para-institutional complexes increased. US domestic opposition to the war and the stretching of US military capabilities across various locations meant that the delegation of force took overt and institutionalized forms. New institutional structures were put in place in Vietnam to administer the outsourcing of the counterinsurgent/unconventional warfare effort. Lessons learnt during the Vietnam conflict paved the way for the concretization of para-institutional complexes in US foreign policy.

US Counterinsurgency Doctrine and the Paramilitary Option

US counterinsurgency training manuals which were used to instruct foreign militaries throughout the early 1950s, 1960s, and beyond advocated delegating central military/security tasks to a collection of para-institutional forces.[14] Counterinsurgent instructional handbooks recommended the active participation of "irregular," "paramilitary," or "friendly guerrilla" forces and "self-defense units."[15] Groups such as "civil defense units," one training manual explained, might include "people from rural areas, ethnic minorities, and other miscellaneous groups such as workmen's militias, youth organizations, and female auxiliaries. They can provide local and internal security of their villages and hamlets when properly trained and armed with adequate weapons." This manual continued to outline how such forces could be trained in "hunter-killer team techniques" and other "guerrilla" tactics such as "raids on insurgent camps" and "intelligence gathering penetrations of insurgent controlled areas."[16] Other documents, such as the US military *Counterguerrilla Operations* (1967), emphasized that "the organization of the paramilitary force may be similar to regular armed forces," yet was to be "constituted from indigenous volunteers" which "include organized youth groups, auxiliary political organizations and part-time, armed civilian militia."[17] Other classified materials such as the *Covert Paramilitary Training Course* (1952) and *The Para-Military Manual Field Handbook* (FM-8000-1, May 1954) provided detailed step-by-step instructions on how to set up pro-government militia forces and paramilitary groups.[18]

Further review of the US counterinsurgency doctrine reveals the policy rationales for mobilizing or backing different types of militias, paramilitaries, and civilian self-defense forces. Such para-institutional forces were generally viewed as efficient and cheap means of mobilizing counterinsurgent forces. In addition, US advisors advocated mobilizing paramilitaries to conduct local policing roles to free up the regular armed forces. Militias of various kinds were understood to confer short-term benefits in the military and political means to defeat an insurgency. However, three rationales stand out. Firstly, informal militias and paramilitary forces provided the plausible deniability to conduct the "dirty work" perceived necessary to eliminate the insurgency and dissolve its civilian support base. By distancing itself from the actions undertaken by militias, the state (and the US) could deny responsibility for them. Secondly, civilians organized into self-defense units ostensibly provided unique advantages in identifying insurgents and their sympathizers. Insurgents and oppositional forces are notoriously difficult for government forces to identify. They live among the population and depend on civilian support. According to US counterinsurgent thinking, local civilian knowledge of the people, their language, culture, and terrain made civilian-based militias important assets for "separating the fish from the sea." Civilian defense forces helped to overcome the problem of identifying those in opposition to the state. Finally, US planners viewed support for paramilitary organizations as a strategy to persuade the members of these forces and the civilian population to actively support the government. US military advisors perceived counterinsurgency to be a population-centric affair that required the active participation of civilians to collaborate in the maintenance and consolidation of a desired political order. Ultimately, local collaboration helped establish some legitimacy for the challenged government.

US military thinkers understood counterinsurgency as a protracted politico-military struggle over the desired political and economic order. The aim was to stabilize "friendly" government control and insulate it from oppositional subversion and dissent "from below." The ultimate objective, as some military manuals describe the desired end state of "stability," was a "national unity" where no substantial ideological, political, or military challenge to the existing state arrangements exists.[19] US counterinsurgent training manuals described how "foreign" ideologies and "attitudes and beliefs" could breed "discontent" and "agitation" among civilian populations.[20] This distinctly political dimension placed civilian activities firmly at the center of counterinsurgent attention. Consequently, unarmed oppositional political and social movements were often portrayed as potentially subversive throughout US counterinsurgent doctrine.[21] US Army FM 7-98 stated "along with overt actions, an insurgency can be characterized by strikes, demonstrations, propaganda, political organisation and diplomacy."[22]

Unions, educational systems, and political parties, to name only a few, were all considered potentially "hostile" elements of society in need of investigation and reorientation. This was especially true, for example, of training manuals distributed to students at the (formerly) School of Americas (SOA), the largest institution in the United States which trained recruits from all over Latin America and the world. One stated that

... it is essential that domestic defense intelligence agencies obtain information about the political party or parties that support the insurgent movement, the quantity of influence that the insurgents exercise, and the presence of the insurgent movement in the non-violent public attacks against the government.[23]

Politically oriented organizations such as unions and other student and social movements were constructed as potentially subversive.[24]

Particularly problematic, according to this doctrine, were attempts to use democratic processes to pursue political change. One manual, *Revolutionary War, Guerrillas and Communist Ideology* (1989), claimed:

The insurgents try to influence the direction, control and authority that is exercised over the nation in general and in the administration of the political system. The insurgents are active in the areas of political nominations, political organizations, political education, and judicial laws. They can resort to subverting the government by means of elections in which the insurgents cause the replacement of an unfriendly government official to one favorable to their cause.[25]

Similarly, a US Army manual entitled *Stability Operations*, stated that political parties associated with the insurgent cause (i.e. socialism or communism) "will attempt to create fronts (or coalitions) of the mass civil organizations to serve the party's interest and gain wide-spread support for its drive to destroy the government." Such organizations consist of "student groups, unions, youth organizations, political parties, professional associations, and possibly religious groups or women's associations. Many of them will have patriotic or democratic names."[26] Statements such as these, prevalent throughout much of the US Cold War counterinsurgency doctrine, reveal the extent to which particular political and social policies, identities, and activities were considered detrimental to the desired order. Perhaps most striking is the endorsement of the suppression of democratic movements in cases where they may yield outcomes at odds with US interests and those of global capital. Counterinsurgency was often tantamount to social and political engineering and often resulted in US support for autocratic and repressive regimes.

Given this understanding of subversion and insurgency and the emphasis on civilian activities as part of a broader plane of social and political contestation, it followed that methods of ridding an area of subversion and gaining compliance would focus on the civil population itself. To do this, US counterinsurgency manuals highlighted Psychological Operations (PSYOPS) (a nice word for propaganda) to eviscerate "subversive" ideologies, the relocation of the civil population to separate them from hard-liners, and the "neutralization," or extermination, of subversive members of society. PSYOPS "encompasses those political, military, economic, and ideological actions planned and conducted to create in neutral or friendly groups the opinions, emotions, attitudes or behavior favorable to the achievement of national objectives."[27] This involved actions to "influence neutral groups and the world community" and to "assist the government in providing psychological rehabilitation for returnees from the subversive insurgent movement."[28] PSYOPS also included physical communicative acts such as civic action (later named Internal Defense and Development, IDAD), which involved handing out food, clothing, medical attention and supplies, and the building of infrastructure, among other activities, in attempts to win over the population and gain their passive and active support for the government and their rejection of the "subversive" ideology. One manual recommended "motivation of population, by such actions as environmental improvements, designed to psychologically condition the population and induce them to participate in the reconstruction of the area and in the defense of their area."[29] The Alliance for Progress, a US economic assistance package and developmental program in Latin America which was often considered a "Marshall Plan" for the region, and funding from the Agency for International Development, for example, might be seen as the public face of counterinsurgency policies.[30]

With the civilian population as the central focus, counterinsurgency combined military campaigns, psychological operations, and civic actions towards suppressing and ultimately defeating both armed insurgency and "radical" political opposition. This often involved stringent control measures in order to extirpate insurgents and active collaborators from a local population. US counterinsurgency strategists advised local military forces to implement programs for the "relocation of those persons of doubtful sympathy" and the resettlement of entire areas.[31] This was described as a form of quarantine where the establishment of "sanitary zones" may be necessary for "rehabilitation" of populations from exposure to subversive ideologies. One manual stated, "In some cases it might be necessary to relocate entire villages" and then it also "may be necessary to relocate those who cannot be protected from guerrilla attack, and those who are hostile and can evade control."[32] The "re-education" of the population was part of this process. Another instructed the forceful removal of the population of an area in order to complete "operations of destruction" of insurgent forces.[33] Direct control over the population was also outlined with little to no reference to the limitations of use of force. This included searches, travel restrictions, rationing, restrictions of imports to an area, curfews, a system of rewards and punishments for certain behaviors and actions, and identity cards, among many others.[34]

US Cold War counterinsurgency strategies also consisted of a series of "unconventional" and "irregular" measures to separate, isolate, and "destroy" or "neutralize" the enemy. For instance, US counterinsurgency planners envisioned the deployment of "hunter-killer teams" intended to "eliminate" guerrillas and, crucially, given the centrality of the civil population, those "underground elements of an irregular force."[35] This formed part of Kennan's vision of "fighting fire with fire" and often involved "counter-terror" to use "guerrilla" tactics against insurgent forces and sympathetic members of the population to deter opposition and subversion.[36] This often involved massive firepower and repressive tactics. Anyone suspected of sympathizing or being involved in subversive activities was a potential target. Some manuals recommended counterinsurgents make "black lists" of the suspected insurgents in the civil population. Others advocated using violence to intimidate civilians into submission. In one declassified document, the US Army delineated "psychological operations" as "communicative acts such as propaganda as well as physical acts of murder, assassination, or a simple show of forces which are intended to influence the minds and behavior of people."[37] Similarly, the US Psychological Warfare Company had a class listed in its program of instruction called "propaganda of the deed."[38] The effective use of "propaganda of the deed" and assassinating key personnel were also included in other manuals.[39] Another counterinsurgency manual warned that

> ... troops employed against irregular forces are subjected to morale and psychological pressures ... and results to a large degree from: (1) The ingrained reluctance of the soldier to take repressive measures against women, children, and old men who usually are active in both overt and covert irregular activities or who must be resettled or concentrated for security reasons.[40]

Yet, this application of force did have limits. One manual stated that counterinsurgent practitioners "may not employ mass counter-terror (as opposed to selective counter-terror) against the civilian population, i.e. genocide is not an alternative."[41]

In direct contradiction to this perceived need to fight "fire with fire" and harsh measures, the US Cold War doctrine also advocated the need to gain the support of the local population (now often popularized by the phrase "winning hearts and minds"). As a politico-military mode of statecraft, the ultimate goal in counterinsurgency was the stabilization of order, rather than just the decimation of enemy forces and oppositional social movements. US counterinsurgent specialists and advisors were often aware of the tension between unleashing coercion to control recalcitrant populations and gaining support for the local government. One manual highlighted this lucidly:

> US forces engaged in counter-guerrilla operations function under restrictions not encountered in other types of warfare. These restrictions may appear to hamper efforts to find and destroy the guerrilla. For example, the safety of non-combatants and the preservation of their property is vitally important to winning them over to the government's side.[42]

Army and Special Force operations manuals highlighted the importance of civil-military affairs and maintaining a positive image for the local host government.[43] Many US military analysts agreed that the use of violence and terror to coerce the local populations (which is described in the counterinsurgency doctrine as an effective tactic undertaken by the opposing guerrilla forces) could be counterproductive and might actually decrease the support for the local government.[44]

US counterinsurgency doctrine contained a tension between the perceived need to militarily eliminate an insurgency, i.e. fighting "fire with fire," and the political objective of "winning the hearts and minds" of the local population. This tension opened the space in which death squad-style informal pro-government militias and paramilitary forces gained their expected utility. While "regular" armed forces were allegedly bound by restraint in order to preserve a positive image and "win hearts and minds," "irregular" forces, such as militias, could operate outside the established norms. This provided the local host government a degree of plausible denial in conducting "dirty war" tactics and allowed for entities informally connected to the state to conduct the harsh tactics and reprisals against populations suspected of opposition, dissent, and subversion. This formed an implicit underlying rationale in US counterinsurgency doctrine for semi-autonomous and informal militias and paramilitary forces. It was not often expressed explicitly in counterinsurgent doctrine, but these lessons were made relatively clear in the formation of what US doctrine referred to as "hunter-killer teams" and "paramilitary forces."[45] The overall lesson, according to French counterinsurgent theorist David Galula, from which the US military has drawn heavily, is that counterinsurgency operations "cannot fail to have unpleasant aspects" and therefore should be undertaken by "professionals" not directly associated with government forces.[46]

Under a slightly different instrumental rationale, US counterinsurgents also instructed their trainees to mobilize a different type of militia via the implementation of "counter-organization"[47] and "consolidation"[48] campaigns, which sought the organization of civilians into "self-defense units" or "civilian self-defense forces." Such civilian-based militias are a recognized category of militia, composed of civilians who operate in their local areas for primarily defensive purposes and to provide information to the government on the identity and location of insurgents. Civilian defense forces were understood as distinct from other paramilitary forces which were more mobile and aggressively sought the destruction of insurgents. US Cold War warriors encouraged the mobilization of the civilian population into "civil defense organizations" to exploit the knowledge and familiarity of local areas, language, and terrain in order to identify, capture, and kill insurgents and their supporters. One US counterinsurgent instructional handbook outlined that militia "agents are recruited among the local residents of the operational area. They have an intimate knowledge of the local populace, conditions, and terrain, and often have prior knowledge of, or connections with members of the irregular force."[49] Similarly, another highlighted the need to use "allied forces" that are "native to the area" under the rationale that "their familiarity with the country, people, language, and customs makes them invaluable."[50] Another stated that "the organization of the paramilitary force may be similar to regular armed forces," yet is to be "constituted from indigenous volunteers whose knowledge of the terrain and people is equal to that of the guerrilla."[51] In playing a central role in the provision of information to their governments and counterinsurgent armed forces, civilian defense forces were often akin to informant networks. For example, one military manual advised deploying "schoolboy patrols" and capitalizing on the information children might have in intelligence-gathering networks.

Outsourcing military and security functions to civilian self-defense forces was also viewed as a politically expedient way to gain their active support for the government. This rationale for such militias is both political and psychological in nature. Civilian militia forces were mobilized under specific rationales such as "counter-organization," and the "mobilization of sympathetic social sectors on the counterinsurgent's behalf" was a "basis through which a neutral—or suspect—population could be regimented and controlled."[52] US Army *Counterguerrilla Operations* instructed, for example, that civilian defense forces may "be organized primarily to indoctrinate their members to support the government."[53] Turning members of the civilian population against the insurgents and those who support the insurgent cause by organizing them into militias was often described as, borrowing from Mao's analogy, "separating the fish from the sea": "The organization and presence of effective local defense units can neutralize the insurgent's efforts to gain support from the people; the insurgent must face the realization that it may now be necessary to fight for support, whereas before, persuasion or threats were sufficient."[54] In effect, the aim of this tactic was to polarize communities by setting in motion a process which would separate them according to two sides: those that are sympathetic to the insurgent cause and those that are sympathetic to the government's. However, the aim in supporting civilian defense force programs was explicitly to reduce civilian passive and active collaboration with the insurgents and increase passive and active collaboration with the government.

According to this logic, then, one way to isolate and extirpate insurgents and oppositional elements of society from the local populace is to organize and persuade civilian organizations to fight on behalf of the government. This rationale for the use of paramilitaries and civilian militias is further explained in a US military article by Paret and Shy:

The ultimate technique in isolating guerrillas from the people is to persuade the people to defend themselves. Militia-type local defense units help in the military defeat of the guerrillas ... But at least as important is their political function: Once a substantial number of members of a community commit violence on behalf of the government, they have gone far toward permanently breaking the tie between that community and the guerrillas.[55]

Robert Trinquier, a classic counterinsurgent theorist, agreed that the most effective way to erode popular support for an insurgency is to mobilize the population in support of the counterinsurgency cause.[56] When implemented, these programs often had a considerable impact on local populations. The inherently political function of civilian defense forces often created the basis for the polarization of entire communities as they are forced to choose one side or another. It also, however, often produced infighting and exposed civilian populations to insurgent reprisals.[57]

US Training Applied and Para-Institutional Forces in Foreign Internal Defense

These lessons were implemented in both US-directed and US-backed counterinsurgencies around the world as part of the broader policies of containment in US Open Door strategies. Dispatched to the counterinsurgency theatre, US agencies—Special Forces and the CIA in particular—trained, assisted, and coordinated allied state armed forces and militias, self-defense forces, and other "irregular forces." The Special Forces' mandate specified their responsibility for "organizing, equipping, training, and directing paramilitary or irregular forces in stability operations."[58] Successive US presidencies in the early Cold War period contributed to the growth of the Special Forces for these purposes. For example, Kennedy's National Security Action Memorandum (NSAM 2 1961) allocated for "an expanded counter-guerrilla program involving in FY 1962 the addition of some 3,000 men to the Army's Special Forces and a budget augmentation of $19 million."[59] NSAM 56 (1961) and NSAM 162 (1962) also aimed to boost US paramilitary capabilities and to link interagency expertise for paramilitary and irregular operations.[60] The then Secretary of Defense Robert S. McNamara was later ordered to re-direct $100 million "to expand and reorient the existing forces for paramilitary and sub-limited or unconventional wars."[61] This also aided the development of overt agency paramilitary operations in US counterinsurgency, such as those of the US Army in Vietnam.

In more indirect affairs, states in receipt of US training and assistance applied these lessons. In this way, US agencies or personnel often did not have direct contact with para-institutional forces abroad collaborating towards internal defense. Instead, US training and military support lent itself to counterinsurgent frameworks in foreign states that did.

Although it is difficult to assess the impact of US counterinsurgency training and financing, they were important in framing the counter-insurgent initiative in particular ways.[62] One manual of the US Army's School of the Americas, *Operaciones de Contraguerrilla*, highlighted the expected influence of military to military ties:

> … the doctrine developed and tested by United States agencies can prove useful in many of the world's nations. The Chief of Mission and brigade commandeers should encourage the military chiefs of the host country to adopt organizations similar to those that have been proven to be efficient in countering guerrilla forces.[63]

The extent of the influence of US doctrine and training is evident in the internal security structures of allied counterinsurgent countries during this time, as discussed below. However, these para-institutional arrangements were not always merely products of US counterinsurgent design. Local agency and processes within military and state develop-ment in the Third World often produced collaborative relationships with para-institutional forces.[64] Local elite interests, corruption, domestic inequalities, and social dislocations were part of important historical developmental patterns that helped generate or at least shape these para-institutional arrangements.

There is a long historical legacy of para-institutional forces in US-supported counterinsurgency wars around the world.[65] For example, as US Army historian Andrew J. Birtle recorded, during the late 1940s and early 1950s, the US provided substantial support to the internal defense of Greece, the Philippines, Indochina, and Korea among other countries. As part of US-backed counterinsurgency in each of these countries, "US advisors … sought to establish defended villages and local self-defense units to free the regular army."

However, Birtle continued, there was "concern over the reputation of paramilitary groups for lawlessness and brutality," which, "led the Army to move cautiously on creating such entities, lest their excesses undermine the goals of pacification." Moreover, he asserted:

> US advisers had very little control over indigenous governments on this score, especially since many governments organized paramili-

tary forces without American material aid. Consequently, the best the United States could do was to urge indigenous authorities to impose tighter control and discipline over the paramilitaries.[66]

In Greece, for example, various local collaborators, militias, and "freedom-loving democrats" were instrumental to the overall US-backed counterinsurgent effort.[67] Alongside substantial US military and economic aid to Greece to build up Greek military capacity, the US Military Advisory Group to Greece served in an advisory role to the Greek government in the containment of communist forces. The US contingent of advisors in Greece established a semi-official village guard system and the mobilization of civilians into a program called the "National Defense Corps." This program replaced previous similar Greek paramilitary militia forces MAD and MAY, which were civilian defense units composed of civilian conscripts mobilized by the Greek Army which fought near their own homes in order to defend their villages from insurgents and to "free up" the regular Army. The National Defense Corps saw its ranks swell to 50,000 as civilians were mobilized during the civil war. The National Defense Corps conducted "governmental guerrilla warfare" and, according to US advisors, was "an instrument of our policy," even though very often neither the Greek government nor US military advisors had complete control over the Corps and were often criticized for their brutality against communist suspects.[68]

The Philippines served as a laboratory for US-supported paramilitary counterinsurgency. As the US granted full independence to the country in 1945, the Philippines stepped up the containment of the Hukbalahap (Huks, for short) rebellion that had intensified due to land ownership inequalities, gradual transformation in rural societies, and food shortages.[69] The US Army and the CIA advised, trained, and in some cases directed the Filipino armed forces in their counterinsurgency to pacify internal dissent and subversion throughout the conflict between 1945 and 1952. The US mainly had an advisory role, but US soldiers themselves sometimes participated in military activities.[70] US advisors in conjunction with the Filipino Army organized and backed various para-institutional forces to help root out the Huks, who had entrenched themselves in remote, mountainous, and jungle terrain. These para-institutional forces included commando paramilitary units, self-defense squads, and former Huk rebels who had switched sides into "friendly" guerrillas.[71] The declassified transcripts of a seminar on the

Huk rebellion and counter-guerrilla operations in the Philippines held at Fort Bragg, North Carolina revealed discussions on how, according to one official, "many forces other than regular troops were used" in the US-backed Filipino counterinsurgency.[72] Some of these paramilitary forces were labeled "Scout Rangers" and "Force X" and "auxiliary" groups, and were trained by US forces to identify and locate Huks and their sympathizers. Filipino leaders, in consultation with US Cold War warriors, such as Edward Landsdale, led elite commando-style hunter killer teams (including one called the "Nenita unit").

The implementation of recommendations for the organization of civilians into "self-defense" units bolstered US-backed para-institutional counterinsurgency networks. US officials described how the "civilian guards" contributed to Filipino counterinsurgent capabilities by soliciting mass local collaboration: "tested civilian guard units were enlarged and given better weapons. Relatives of civilian guards became potent anti-Huk spies in spite of lack of instruction. The whole operation snow-balled into entire communities backing up their own home guard organizations against the Huks."[73] In addition, the Filipino government authorized large landowners and rural businesses to set up "private armies" or local security forces to protect their property and businesses against the Huks' advances.[74] These programs and other similar paramilitary efforts were precursors to the militias, and vigilante and paramilitary groups that spearheaded the US-backed counterinsurgency in the Philippines during the 1970s through the 1990s.[75]

In Guatemala, after the US orchestrated the overthrow of Arbenz, US advisors and US military assistance helped support a counterinsurgent apparatus that would go on to mobilize hundreds of thousands of civilians into vigilante self-defense forces and to deploy pro-government militias and death squad-style paramilitaries against leftist guerillas, political parties, unions, and civil movements sympathetic to their cause.[76] Guatemalan counterinsurgency, as numerous declassified US documents attest to, involved methods that terrorized civilian populations into submission, often delegating extreme violence to death-squad actors.[77] These programs were often backed by the US, were morally condoned, or at the very best, ignored. For example, in one declassified 1968 US diplomatic cable, Viron Vaky (Deputy Chief of Mission to Guatemala 1964–67 and later assistant Secretary of State for Inter-American Affairs) lamented that the US, in supporting Guatemalan internal defense, has "condoned counter-terror," and had broadly sent the message that "murder, torture, and mutilation are alright if our side is doing it and the victims are communists."[78] The problem with this, he argued, was not merely a "practical political" one due to the public image of US intimate involvement in these forms of state terror, but also a "moral" issue that could ultimately have threatened the effectiveness of the counterinsurgency drive. In other messages, he deplored US involvement in Guatemala's indiscriminate "counter-terror" to combat the insurgency and its use of death squad-style militias.[79]

During the 1980s, the Guatemalan Armed Forces organized thousands of civilians into "civilian self-defense patrols" (PACs) according to US counterinsurgency doctrinal recommendations.[80] Although participation in these pro-government militias was meant to be voluntary, many people were violently and forcibly recruited. Many were made to attack their neighbors and areas suspected of communist infiltration.[81] By 1984, according to some estimates, the PACs had recruited over 900,000 civilians, primarily men from indigenous ethnic backgrounds and lower classes, to actively participate in both defensive and offensive counterinsurgency measures.[82] Numerous studies, such as the Guatemalan truth commission report *Memory of Silence*, show how such para-institutional forces, particularly in the early 1980s, were responsible for terrorizing local populations suspected of being communist or sympathizing with the communist cause.[83]

In El Salvador, according to extensive documentation, the introduction of US counterinsurgency training and assistance, to protect US political and economic interests and to "stabilize" favorable state formations, had a noticeable effect on the structure and organization of El Salvador's security apparatus.[84] American military advisors, the CIA, and Special Forces arrived in early 1961 and played an intimate role in the creation of para-institutional formations. In 1963, the CIA and ten Special Force advisors aided Salvadoran General Jose Alberto Medrano in the creation of the *Organización Nacionalista Democratica* (better known by its acronym ORDEN, symbolically meaning "order") with the restructuring of the entire military apparatus, even before a significant oppositional armed insurgency was mounted.[85] ORDEN, as CIA records indicated, was "comprised of tens of thousands of conservative rural peasants as an intelligence gathering organization—identifying and taking direct action against real and suspected enemies of the regime."[86] It depended almost entirely on civilians in rural areas, but drew on the expertise of ex-military personnel and military reserves.

As ORDEN grew, however, this program and others under US guidance produced

> ... a vast network of paramilitary irregulars feeding information into the intelligence apparatus, providing manpower for counter-insurgency's dirty work, and serving as a back-up army of irregular auxiliaries to be activated for large-scale security operations whenever the need arose.[87]

These "anti-guerrilla forces, specially trained, utilizing qualified local men, are much more economical in cost, number, and results than large forces using conventional methods," according to Colonel Rodriguez of the El Salvadoran Army.[88] The El Salvadoran state and armed forces, with US guidance, also established ANSESAL (*Agencia Nacional de Seguridad de El Salvador*—the National Security Agency of El Salvador), another semi-official paramilitary agency connected directly to the presidency.[89] These programs formed the basis for the death squads operational during the 1980s, which later became familiarly known as the "Salvador Option."[90]

ORDEN, ANSESAL and the development of other death squad-style actors were not entirely an American fabrication for the fulfillment of US geopolitical objectives. These were products of the influence of US counterinsurgent programs within the context of domestic dynamics and political and economic tensions that had fractured society towards internal conflict. El Salvador's history since independence had roughly been characterized by revolution and counter-revolution in which the privileged minority clung to their powerful positions against the poor majority by means of military repression.[91] El Salvador's historical con-testation of political space in which the elite social forces saw it necessary to suppress peasant uprisings and popular calls for reform had already led to a strong tradition of military social control. As a 1981 CIA official reflected in one report, "The ultra-right in El Salvador has a long history of using violence as a political tool, perhaps marked most vividly by the widespread repression and murder of *campesinos* following the failed peasant rebellion in 1932."[92] Wealthy land owners and elite political strata in control of the political system financed vigilante paramilitary organizations as personal security guards to protect assets threatened by peasant movements, the land reforms of the 1950s, and the communist political agenda.[93] In conjunction with these local interests, the refraction of El Salvador's political problems through a Cold War lens arose from domestic interpretations of international politics as much as one of a US imposition. In short, the implementation of paramilitary networks within El Salvador's counterinsurgency state was inseparably embedded in domestic political and economic structures.

Yet, these internal dynamics in the landscape of El Salvadoran contentious politics do not detract from the impact of US security training and the framing of El Salvador's structural issues as one of a Cold War problematic. Nor do they negate the para-institutional frameworks upon which US-backed counterinsurgency efforts resided. The broader picture is one in which para-institutional formations functioned in the maintenance or stabilization of state arrangements conducive to a US-driven liberal order. Although US officials have frequently denied direct connections to death-squad agents, declassified documents demonstrate that the containment and elimination of left-wing "subversives" was prioritized over preventing paramilitary structures from operating as death squads.[94]

The formation of paramilitary organizations in Colombia also had firm roots in US counterinsurgency training and doctrine. Training programs at the SOA, US Army Special Warfare School, and at the Lancero Military School of Instruction (a training center set up by US Special Forces in Colombia), and substantial US military financing formed the basis for a para-institutionalization of Colombia's counterinsurgency efforts.[95] In a classified supplement attached to a 1962 report, a Special Force team led by General Yarborough of the US Special Warfare Centre recommended that a

> …concerted country team effort should be made now to select civilian and military personnel for clandestine training in resistance operations in case they are needed later … This structure should be used to pressure toward reforms known to be needed, perform counter-agent and counter-propaganda functions and as necessary execute paramilitary, sabotage and/or terrorist activities against known communist proponents. It should be backed by the United States.[96]

According to Dennis Rempe, writing in 1995, "Owing to the sensitive nature of Colombian internal security missions, the survey team further advised the use of third country nationals, covertly under US control, but apparently contracted by the host government."[97]

The introduction of the US counterinsurgency doctrine had significant influence on the Colombian military understanding of the problems of insurgency and how to counter it, including the mobilization of paramilitary forces.[98] One Colombian Army field manual, *Operaciones Contra Las Fuerzas Irregulares*, was a translation of the US Army's FM 31-15 (as analyzed above) with the same title in English. *Reglamento de Combate de Contraguerrillera* references numerous US counterinsurgency texts and others contain very similar content.[99] Much like in the US doctrine, many civilian activities were considered potentially "subversive" and advocated the need to respond with "counter-organization" and stipulated that "self-defense" systems should be planned for the "violent rejection of guerrilla actions in their region."[100] For instance, one Colombian counterinsurgency handbook dated 1963 illustrated that "both sides must force the local population to participate in combat; to a certain extent the inhabitant is converted into a combatant." Another 1962 manual instructed the reader on "the obligation of part of the inhabitants to participate in their own defense." Similarly, *Reglamento de Combate* advocated the need to "organize the civil population militarily, so that it can protect itself against the actions of the guerrillas and support combat operations."[101]

The Colombian government, with US backing, created militia and paramilitary programs. Decree 3398, announced as a part of a declared "state of siege" during the US-supported program called "*Plan Lazo*" (1962–65), laid the legal basis for arming of civilians and their incorporation into the counterinsurgency effort. It granted the army the recourse for "the organization and tasking of all of the residents of the country and its natural resources ... to guarantee National Independence and institutional stability."[102] Then Law 48 (1968) legalized civilian militias and advocated the utilization of civilians for the "private use of the Armed Forces."[103] Article 183 of resolution 005 of April 9, 1969 also legalized "organizing in military form the civilian population."[104] Not surprisingly, members of the US State Department were "delighted to hear this declaration for the determination of a number of leaders in Colombian life to halt the spread of communist ideas."[105] Paramilitary forces became commonplace in Colombia during the early Cold War and expanded significantly throughout the 1980s and 1990s with deadly consequences for those suspected of associating with communism, or progressive, left-wing social movements. However, as will be discussed further in Chapter 5, the evolution of paramilitarism in Colombia has been traced to the protection of large-landowner interests, multi-national companies, and conservative political elite.[106] Paramilitary forces formed in the interaction between the US promotion of certain counterinsurgent practices and domestic dynamics of social and political violence.

These examples of US-backed foreign internal defense across Central and South America and South East Asia help to demonstrate the mostly indirect relationships that comprised US imperial paramilitary forces in counterinsurgency settings. US training, doctrine, and support had powerful indirect effects on recipient state military structures and significantly contributed to the development of para-institutional forces. In other cases, US personnel worked more behind the scenes in client-state military institutions or directly organized the formation of para-institutional forces themselves. However, local elite agency and domestic structural political and economic dynamics cannot be ignored. The intersection of transnational capital interests and local elite activity shaped the development of para-institutional complexes. The overall effect was that US imperialism, working to keep countries "open" to international markets and investments, was underpinned by expansive para-institutional forces. Similar patterns and relationships were observed in many other parts of the world.

Vietnam and the Evolution of the US Use of Imperial Para-Institutional Complexes

US intervention in Vietnam marked an important point in the development of US-backed para-institutional networks. US advisors relied on various para-institutional forces, including militias, paramilitary groups, mercenaries, and PMCs to a greater extent than ever before, for a variety of different reasons. With US direct intervention and conventional military forces on the ground in Vietnam (1965–73), outsourcing became overt and institutionalized. Military and civilian agencies, such as the Army and USAID (as opposed to the covert Special Forces and CIA) devised institutional frameworks for the administration of para-institutional forces. This helped to cement practices of outsourcing into US foreign policy. Simultaneously, broader pressures in US global capabilities pushed for further outsourcing of US statecraft abroad. Mired in multiple engagements, outsourcing offered an alternative means to extend US coercive reach. This was compounded by domestic opposition within the US to the Vietnam War (eventually culminating in a "Vietnam syndrome") and Congress-imposed caps on troop numbers towards the end of the conflict, which meant building alternative ways to continue US efforts via paramilitary assets, and hiring private military services and third-country mercenaries (the "More Flags" program).

US involvement in Vietnam and Laos throughout the late 1950s and early 1960s was consistent with policy priorities to "remain in the background." Kennedy, for example, repeatedly rejected requests to send US troops to support the South Vietnamese government (officially the Republic of Vietnam) and its Army of the Republic of Vietnam (ARVN) against a growing insurgency and their northern communist neighbors. Instead, US policymakers authorized hundreds of millions of dollars in counterinsurgency aid (1953–63) and CIA and Special Forces training and covert actions. The Kennedy administration also drafted a plan in early 1961 to supplement the South Vietnamese combat efforts with private companies and counterinsurgency and paramilitary training centers.[107]

In the early 1960s, US advisors intensified US-backed counter-insurgent paramilitary programs. US advisors organized civilians into "self-defense" units via the Strategic Hamlet Program. In official assess-ments of US efforts in Vietnam and the hamlet militias, US advisors extensively described the utility and concepts for arming and training civilians for the "pacification" of dissent and "winning the peasants" over to support the Southern Vietnamese government.[108] US military reports estimated that by 1963 "approximately 11,000 strike force and 40,000 hamlet militia from over 800 villages had undergone training that averaged about six weeks for strike force troops and two weeks for hamlet militia."[109] According to another figure, in 1964, these self-defense militias comprised of 196,000 armed civilians, nearly equaling South Vietnam's regular army of 250,000 personnel.[110] US advisors also created the Civilian Irregular Defense Groups (CIDG) program in early 1961. The CIDG program saw the Special Forces (in coordination with the CIA) hire tens of thousands of Vietnamese minorities, primarily amongst the ethnic hill tribes colloquially called the Montagnards, as well as ethnic Cambodians and Chinese, many of whom were trained and paid as militias and mobile strike teams (also known as "hunter-killers"). An official US Army report recorded that these forces would be organized around "guerrilla-style" tactics, trained specifically in "ambushing, raiding, sabotaging and committing acts of terrorism against known VC (Viet Cong) personnel."[111]

As US support for and involvement in South Vietnam's counter-insurgency intensified, the institutional management of expanding paramilitary programs were transferred to overt US government agencies. Kennedy's National Security Action Memorandums 55, 56, and 57 assigned responsibility for paramilitary operations to the US Armed Forces, rather than its more covert intelligence counterparts. NSAM 57, for instance, titled "Responsibility for Paramilitary Operations" stated that

Where such an operation is to be wholly covert or disavowable, it may be assigned to CIA, provided that it is within the normal capabilities of the agency. Any large paramilitary operation wholly or partly covert which requires significant numbers of military trained personnel, amounts to military equipment which exceed normal CIA-controlled stocks and/or military experience of a kind and level peculiar to the Armed services is properly the primary responsibility of the Department of Defense with the CIA in a supporting role.[112]

This move to overt agencies signaled a broader shift in bringing US paramilitary programs into the conventional military fold, but also had implications for Vietnam. The Military Assistance Command, Vietnam (MACV) was created in 1962 under the US Army. The MACV took control of the CIDG program in 1963, as part of Operation Switchback.

US involvement escalated, and in 1965 American troops arrived in Vietnam. With the US military presence, military advisors increased the para-institutionalization of counterinsurgent efforts. Under the direction of the MACV, the CIDG became further offensive in nature, often with US forces fighting alongside these for-hire militia armies. According to Douglas Blaufarb, a former CIA officer, "the armed tribal irregulars [the CIDG], were no longer a hamlet militia ... They were used for attack and defense against enemy units," and in this role were "close to being mercenaries."[113] Much like many other paramilitary operations in Laos, Cambodia, Tibet, and elsewhere, these forces were supported by Air America (and sometimes the US Air Force) under contracts with the CIA and the DoD to run supplies and ferry the Montagnard fighters from one village to the next, identifying and "neutralizing," or capturing, suspected VC and their sympathizers. By 1967, one estimate put the numbers at 2,726 US Special Force advisors in liaison with 34,300 CIDG, 18,200 regional mercenary forces, and about 5,700 mobile strike teams ("hunter-killers").[114] Similar paramilitary programs were initiated with various other tribal groups such as the Sedang, Hre, and Bahnar, amongst others.[115] These paramilitary organizations played a significant and overt role in the Vietnam War as the entire country became engulfed in conflict.

Much like the CIDG, the Phoenix program and its Provincial Reconnaissance Units (PRUs) component, a related and more controversial paramilitary program, was handed over to the MACV in 1968–69.[116] The Phoenix program trained Vietnamese nationalists to locate and "neutralize" suspected Viet Cong (VC) suspects and essentially functioned as a contracted mercenary outfit.[117] To facilitate this program and its intelligence requirements, the CIA and the US Army hired Pacific Architects and Engineers in 1964 to construct interrogation facilities across South Vietnam.[118] The Phoenix program was often referred to as an "assassination program" in the press, with large numbers of suspected VC killed (from 1968 (2,229) to 1969 (4,832)).[119] The CIA website (2011) described how the PRUs "went to the villages and hamlets and attempted to identify the named individuals and 'neutralize' them. Those on a list were arrested or captured for interrogation, or if they resisted, they were killed."[120] One reporter, in an article titled "The CIA's Hired Killers," wrote that the PRUs were "the best killers in Vietnam," stating that they were not much different from the "terrorists" the US was seeking to defeat, except that rather than ideology, they terrorized for money.[121] The Senate Foreign Relations Committee held a hearing on US pacification programs in Vietnam in 1970 in order to unveil the level of atrocities.[122]

The policy rationales for mobilizing civilians into militias, supporting paramilitary formations, and hiring mercenary armies were consistent with US doctrinal recommendations. The Strategic Hamlet Program, following US counterinsurgent logic represented a politico-military strategy to gain the allegiance of a group of people by making them targets of the enemy, and thereby forcing self-defense members to fight back on the government's behalf. US policymakers also viewed all of these programs as an expedient way to free up regular troops (in this case, the Army of the Republic of Vietnam—ARVN) for counter-guerrilla warfare. Another perceived advantage of delegating central tasks to militias and paramilitary assets in Vietnam was that they were cheap. According to a US government estimate, it was up to ten times less expensive to hire a local "peasant" warrior than it was to support a US soldier; a difference well celebrated by US planners.[123] Moreover, local assets were expendable: the death of a local tribesman was not as politically sensitive for US policymakers as the death of a US soldier.

In order to further augment the war capability whilst simultaneously decrease the burden on US soldiers, and thereby mitigate the political consequences associated with large troop deployments, Kennedy's NSAM 162 called for the increased use of "third country personnel." According to the memorandum, "Such forces would be composed of foreign volunteers supported and controlled by the US." Later, as part of the Phoenix program, President Lyndon B. Johnson contracted a "Filipino Civic Action Team" for $39 million to combat the VC and their political following in Tay Ninh province.[124] Then, in 1964, President Johnson began the "More Flags" program in which a variety of third-country troops were hired.[125] The program was initiated under the original objective of providing coalition (i.e. "more flags") non-combat assistance to South Vietnam, but soon expanded to delegate military operations. The Australian and New Zealand forces were sent on a voluntary basis, but the Korean (providing around 50,000 soldiers), Philippine, and Thai contingents were paid for; this included deployment costs, a per diem payment, overseas allowance and death benefits, and cost the US tens of millions of dollars. The Johnson and Nixon administrations went to great lengths to keep the payments to these third countries secret. Around 5,000–6,000 of these foreign national mercenaries died in Vietnam. Congress attempted to restrict the use of the "More Flags" mercenaries, and in 1970 placed an "anti-mercenary" provision in the Cooper-Church Amendment, which sought also to restrict US troops and advisors from operating in Laos and Cambodia.[126]

In a similar fashion to self-defense units, militias, and mercenaries, the use of private contractors gradually shifted from being the preserve of covert operatives to conventional US military and government practice. US Army and Air Force contracted CAT to help resupply French troops across the Indochina region in the 1950s towards French counterinsurgency in phasing out its colonial rule over the country.[127] William Leary noted: "While reluctant to commit American military personnel to the war in Indochina, the Eisenhower administration was anxious to assist the French. This led to a decision to use CAT pilots to fly an airlift in US Air Force-supplied C-119s."[128] Some portions of the training of South Vietnam's forces was contracted out, in one case to a Michigan State University Group.[129] During US direct involvement in Vietnam (i.e. troops on the ground) from 1965 through 1973, private companies were hired by the CIA, US Air Force, US Army, US Agency for International Development (USAID), and State Department, amongst other US government agencies, to provide transportation for US and South Vietnamese government personnel and armed forces as well as evacuation, airlift and supply missions, and other supporting and central combat roles.[130] To manage these arrangements, the Army created an ad hoc administrative body, the United States Army Procurement Agency Vietnam, spending around $500 million on contracts at the peak of the Vietnam War in 1968, and hiring over 50,000 contractors.[131] In total, over 80,000 individual contractors contributed to US operations during US presence in Vietnam.[132] PMCs represented such a crucial component of the Vietnam effort that *Business Week* described it as a "war by contract."[133]

US planners increasingly hired PMCs for logistical purposes to support a large US military presence (1965–73) that surpassed that of previous engagements: "Never before had the Army's logistic system been tasked with the mission of supporting large numbers of ground combat troops operating in a counter-guerrilla role."[134] Many activities which the US military had traditionally performed itself were contracted out to private companies, such as utilities, repair services, base construction, and the procurement and distribution of resources. Vinnell won contracts worth hundreds of millions of dollars for logistical and technical services rendered to support the US military, with around 5,000 contractors on the ground at the height of its involvement.[135] All of the US armed forces relied heavily on Pacific Architects and Engineers, Brown and Root, and other contractors for substantial construction projects and engineering works such as building bases, airports, utilities, hospitals, bridges, among others.[136] Multiple US agencies hired CAT and AA as the principal support mechanisms for the CIDG and other US-backed paramilitary programs across Vietnam and Laos. These airlines carried weapons and supplies, transported fighters, provided evacuation services, and occasionally were involved in aerial combat and bombing raids.[137]

After 1965, AA received a wide assortment of contracts. The majority of such logistical contracts were set by USAID to transport materials for development programs and civic action. The United States Army Support Group, Vietnam (later designated United States Army Support Command, Vietnam) contracted AA for various transport and rescue missions. The DoD hired AA to facilitate research on defoliants and communication infrastructure, including the use of Agent Orange. AA was also involved in the transportation of North Vietnamese prisoners to interrogation centers as part of the Phoenix program, including to the island prison at Côn Sơn where political prisoners were held in harsh conditions.[138] In sum, AA's presence was much more conspicuous than in covert paramilitary operations in other parts of the world during the 1950s and 1960s; even Miss America was flown around Vietnam on a publicity tour campaign in an AA plane.

PMCs were also often contracted as military trainers and mercenary fighters. For example, Vinnell was contracted to run "black" operations, with one Pentagon official interviewed by the *Village Voice* describing them as "our own little mercenary army in Vietnam ... we used them to do things we either didn't have the manpower to do ourselves, or because of legal problems."[139] Contractors helped train South Vietnamese forces. The CIA contracted out many of its activities in liaising with the para-military PRUs and the Phoenix program, usually to former US military personnel.[140] The CIA also reportedly hired Cuban contract agents from the Bay of Pigs mission to fight under the Phoenix program.[141]

Largely in response to growing domestic opposition to the war, US forces were phased out in accordance with Nixon's policy of "Vietnam-ization" (1969–73). The withdrawal process depended on strengthening the South Vietnamese military and government, but also to a large degree on a collection of private companies and paid paramilitary assets that fought at the forefront of the conflict.[142] Contractors also took on new roles, often replacing US agencies altogether. For example, Computer Science Services Inc., under contract from the DoD, took over handling and organizing intelligence of South Vietnamese counterinsurgent agencies.[143] While some contracts, such as with USAID to supply materials for the CORDS program (Civil Operations and Revolution-ary Developmental Support) and development initiatives expanded, AA became increasingly involved as an evacuation service and in the trans-portation of refugees.[144] AA was also commissioned to train Vietnamese pilots.[145] The DoD increasingly depended on public-private partner-ships to extend US influence throughout the US troop withdrawal process. DoD-AA contracts climbed from $17.2 million in 1972 to $41.4 million in 1973.[146] Unbeknownst to many, the famous image of civilians scrambling to board a helicopter parked on a roof in Saigon in 1975 is of an Air America helicopter, not of a US military one.

Overall, several factors played a role in this increased reliance on contractors to support US military deployments. According to a US Army study by Lieutenant General Joseph M. Heiser, Jr., limitations on the deployment of US troops was one of the principal factors in the greater use of contractors.[147] The number of US troops to be deployed to Vietnam after the escalation of war in 1964 was always less than what the MACV requested, which led to an inadequate logistical support base.[148] PMCs filled the gaps, freeing up regular solders for training, combat, and other roles. Similarly, on an international level, part of the pressure towards the privatization of force in Vietnam and Laos came from the over-commitment of US forces around the world. This included covert agencies. Declassified official history of the CIA war in Laos revealed that in 1961, the US Air Force was forced to look into contracting private companies because CIA contractors were committed in so many other places, such as in Cuba for the Bay of Pigs invasion.[149] PMCs augmented military capacity as "force multipliers."[150] In addition to this, contracting offered a way to distance the US from the actions of contractors taken on its behalf. As Dickinson asserts, "the mere fact that these foreign fighting forces were not literally US troops helped the US government distance itself from their actions, rendering abuses more likely, and legal and democratic checks less so. This is a pattern that has continued to the present."[151] Similarly, while the deaths of US troops would not be tolerated at home, those of local agents, mercenary forces, and private contractors largely went unnoticed. Outsourcing represented a way to ameliorate the political costs of imperial war.

Conclusion

As part of the US's Open Door grand strategy to stabilize state arrangements abroad favorable to global capital interests, the US provided substantial military training and assistance towards the internal defense of foreign states, particularly in the global South (or Third World). In many cases, US agencies covertly created and/or supported militias, paramilitary forces, and mercenaries. More indirectly, training programs imparted to foreign state militaries also advocated the mobilization of such groups. Military planners often hired contractors to facilitate or augment these modes of US influence in civil conflicts and contested politics abroad. In this way, para-institutional forces were instrumental to US counterinsurgent statecraft and by extension to the stabilization of state arrangements geared towards US interests.

As the next chapter elaborates, Vietnam set a precedent for the increased use of para-institutional forces in US foreign policy. Moreover, the 1970s context made it difficult to execute US Open Door strategies. Congressional reviews into the CIA's activities, such as the Church Committee (1975), limitations on presidential autonomy, and increased oversight, and domestic opposition to US involvement in far-away conflicts in the form of a "Vietnam syndrome" all inhibited direct intervention. Subsequently, Washington sought para-extended means to police the periphery. Similarly, contracting to PMCs became the norm rather than the exception during Vietnam, and provided the basis for their expanded use elsewhere.[152] US policymakers devised new institutional mechanisms for managing and contracting out military and security tasks. Training of foreign forces (or authorized US PMCs to contract with foreign countries), for instance, such as Vinnell's training of the Saudi National Guard during the 1970s and beyond, became more prevalent.[153]

4
Reagan, Low-Intensity Conflict, and the Expansion of Para-Institutional Statecraft

The Reagan administration expanded the use of para-institutional means to engineer politics in countries in the global South. Outsourcing, contracting, and support for a variety of para-institutional forces during the 1980s became a primary extensional coercive means of influence over the direction of political events in strategic areas. Washington institutionalized many of these practices. This provided the basis for their further development in the 1990s and later in the "War on Terror." In the 1980s, US military planners augmented Special Forces', the CIA's, and other agencies' capacity for paramilitary actions under the revised umbrella rubric of "low-intensity conflict." In addition, during the Reagan presidency, contracting to PMCs was also expanded and institutionalized as a common, overt military practice. The CIA fully privatized its semi-private air-contractor proprietaries and new, modern forms of PMCs developed in their wake. Policymakers in Washington also devised the institutions and procedural infrastructure to manage military privatization, which provided the basic frameworks for contracting in the future. This ensured that the "burden of low-intensity conflict would fall on allied foreign forces, proxies, or mercenaries."[1]

In doing this, Reagan also made support for outsourcing and para-institutional forces an overt and public component of US foreign policy. While previous para-extensional engagements were largely conducted covertly by the CIA and Special Forces, following their successful implementation in Vietnam, Reagan made visible his unabashed backing for para-institutional means of extending US influence. Reagan invested heavily in public relations to help inculcate a prominent "freedom fighter" discourse, which not only aimed at gaining public and Congressional support for these activities, but also to leverage the participation of "private" groups to implement them. Yet, somewhat paradoxically, while Reagan publicly endorsed supporting "freedom fighters" and encouraged non-state participation in US-led low-intensity conflicts, the actual practice of these wars were often driven further to conceal US complicity in actions taken on its behalf.

Domestic and international developments aided in the expansion of these practices in US foreign policy. The US's managerial role in the international system remained intact, but increased domestic Congressional scrutiny over US military affairs and a "Vietnam syndrome" led to greater pressures to outsource global commitments. US planners also regarded outsourcing during Vietnam as a success and applied these lessons elsewhere. The US Army and other agencies believed contracting and leveraging local para-institutional armed forces was an efficient and cost-effective means of projecting US force. A series of measures were introduced to institutionalize lessons garnered from the experience of outsourcing in Vietnam. Finally, as neoliberal capitalism gained traction under Reagan, emphasis on privatization and the efficiency of market mechanisms boomed.

The "Second Cold War" Context

The Reagan administration (1981–89), in what came to be known as the "Reagan Doctrine," renewed US commitments to bolster the advancement and maintenance of a US-led order under the rationale of stemming the rise of Soviet influence and communism abroad. This period is sometimes referred to as a "second Cold War."[2] To do this, Reagan fortified the twin policies of "containment" (counterinsurgency) and "rollback" (unconventional warfare) repackaged under a new umbrella title of "low-intensity conflict." Devised and articulated through a series of conferences and military committees on the subject, "low-intensity conflict" was promoted as a new politico-military strategy to reverse challenges put forward by counter-hegemonic forces.[3] However, instead of offering a profoundly altered understanding of the problems of dissent, subversion, insurgency, and terrorism, as well as responses to them, low-intensity conflict represented a continuation of Cold War counterinsurgency and unconventional warfare. The concepts that underpinned low-intensity conflict were consistent with early Cold War counterinsurgency and unconventional warfare. The doctrine was virtually identical.

Much like earlier Cold War US interventionism, this involved the suppression of local social and political forces in the global South threatening to steer their countries onto alternative developmental pathways. It also included the "overthrow of governments that seek full independence from the economic, political or military influence of the United States."[4] In this manner, "low-intensity" forms of statecraft were consistent with US imperial strategies, with the primary focus on curbing the tide of revolutionary movements around the world. Beyond containing communism and Soviet and Cuban influence, US military planners understood that low-intensity conflict comprised methods to counter obstacles to the fluid flow of global capital transnationally and to integrate countries into the US-led liberalized global order. Economic and geo-strategic imperatives were integral to the broader structural frameworks that guided low-intensity conflict policy.

Official US military statements and low-intensity conflict documents reflected recognition of the ultimate aims of low-intensity modes of statecraft. For example, according to Reagan's 1987 *National Security Strategy*, the architects of the "low-intensity" strategy were fully cognizant that *losing* low-intensity conflicts could lead to the "Interruption of Western access to vital resources ... Gradual loss of U.S. military basing and access rights ... Expanded threats to key sea lines of communication ... Gradual shifting of allies and trading partners away from the United States into positions of accommodation with hostile interests." Therefore, the document pointed out, "If we can protect our own security, and maintain an environment of reasonable stability and open trade and communication throughout the Third World, political, economic, and social forces will eventually work to our advantage."[5] This should be accomplished, it continued, using political and economic inducements designed to "reduce the underlying causes of instability of the Third World, help undermine the attractiveness of totalitarian regimes, and eventually lead to conditions favorable to US and Western interests." Crucially this approach should be coupled with "indirect rather than direct applications of military power."[6]

In an article in *Military Review,* US Colonel James argued more bluntly in favor of a broad stratagem to "better influence politico-military outcomes in the resource-rich and strategically located Third World area."[7] Promoting "openness" in Third World states for US and Western interests, it was argued, was central to US pursuit of mutually reinforcing security and economic objectives. In a similar manner, US military training manuals of the time connected these currents in US foreign policy to low-intensity conflict. FM 100-20 *Military Operations in Low Intensity*

Conflict elaborated that "unfavorable outcomes of LIC [low-intensity conflict] may gradually isolate the United States, its allies, and its global trading partners from each other and from the world community." Losing low-intensity politico-military struggles threatens "the loss of US access to strategic energy reserves and other natural resources," whereas "conversely, successful LIC operations, consistent with US interests and laws, can advance US international goals such as the growth of freedom, democratic institutions and free market economies."[8]

The US's global commitments as part of its Open Door strategy remained intact; however, a "Vietnam syndrome" and mounting Congressional restrictions in US foreign policy activities led to greater pressures to outsource much of this managerial role. After Vietnam, significant domestic opposition to US troop commitments abroad made justifying US interventionism into the Third World to a war-weary public more difficult. Simmons argued that the development and popularity of low-intensity conflict was itself a "principal consequence of the Vietnam Syndrome" and with it indirect proxy methods of implementing the Reagan Doctrine, "where the level of the US personnel involvement was sufficiently low to be disguised for domestic consumption."[9] As part of this, US military planners looked for outsourced means to continue US global foreign internal policing commitments. As Molloy put it, "US intervention is continually masked to avoid potential US domestic opposition by deliberately 'privatising' and 'civilianising' US actions."[10] The *raison d'être* of low-intensity conflict was to avoid direct US military engagement and refrain from the use of US personnel in combat roles.[11] As Secretary of Defense Weinberger argued, low-intensity forms of conflict were necessary "to project United States power where the use of conventional forces would be premature, inappropriate, or infeasible."[12] This indirect approach made it possible to sustain military campaigns in multiple theaters in various locations without the need for public support or approval. Low-intensity conflict, in other words, was designed specifically to avoid direct military intervention, yet could be conducted so as to achieve the desired political and economic objectives abroad.

In addition, Congressional investigations, such as the "Church Committee," and restrictions on US militarism were mounted between 1970 and 1980. Some of the most prominent examples included the War Powers Act of November 1973, which gave Congress more power in controlling the deployment of the armed forces, and the 1974 Hughes-Ryan Amendment, and the 1976 Clark Amendment, which both prohibited

US military assistance to para-institutional forces in Angola. "Since the Vietnam War," a Reagan NSC staff member expressed to a reporter, "we have had this growing involvement by the legislative branch in the details of foreign policy that—you can make a constitutional argument— are properly left to the president. When you do that, you drive him in the direction of using other techniques to achieve objectives."[13] The case of the Contra war against Nicaragua revealed how far Washington was prepared to go to continue its global counter-revolution. As Grandin noted:

> It was in Central America that unconventional warriors learned to bypass Congressional oversight by creating a semiprivate, international network to carry out a clandestine foreign policy and to undermine post-Vietnam efforts to limit the use of military power for other than clearly defined, limited objectives.[14]

Expanding the Paramilitary Option

Through this commitment to low-intensity conflict, Washington expanded and consolidated its reliance on para-institutional forces. US military leaders with experience in Vietnam applied the lessons they had learnt in mobilizing local militia groups. For example, paramilitary programs in El Salvador's counterinsurgency implemented during the 1980s, such as the development of the civil defense programs, were modeled on lessons taken from Vietnam.[15] US advisors also applied lessons from Vietnam in the Philippines.[16] To implement such programs, policymakers in Washington built up the Special Forces and enhanced the CIA's paramilitary capabilities to maximize the use of foreign militia and paramilitary forces, mercenaries, and private companies.[17] The number of Special Forces personnel surpassed previous figures: from a peak of 13,000 active duty officers in 1969, numbers surged from 11,600 in 1981, to 14,900 in 1985, reaching around 20,900 by 1990. Accordingly, the DoD budget allocations for Special Forces increased from around $500 million in 1981 to around $1.2 billion in 1987. CIA covert operations, including paramilitary campaigns, increased five-fold during Reagan's first term.[18]

Respectively, there was a considerable increase in the support for localized para-institutional forces for unconventional warfare and counterinsurgency. In the words of Special Operations Chief Colonel Roger Pezzelle, Special Force mobile training teams (MTTs) and the CIA were dispatched around the world to liaise with "host country regular units, militia, reserve forces and security units."[19] Shortly after assuming the presidency, Reagan ordered various directives to wage unconventional warfare to "roll back" unwanted governments. By 1986, the US, with a budget of over $600 million, was financing "freedom fighters" in four countries. This included support for an estimated 150,000 irregular fighters in Afghanistan, 25,000 in Angola, 20,000 in Kampuchea, and 15,000 in Nicaragua, and sponsoring or supporting similar unconventional warfare programs in Yemen, Cambodia, Chad, Libya, Iraq, and Grenada, among others.[20]

There was a gradual escalation of US backing for para-institutional forces across different cases during the 1980s. For example, Reagan repealed the 1976 Clark Amendment that prohibited military training to the insurgency in Angola, which the CIA had reportedly violated anyway.[21] US funding for UNITA forces increased from $15 million a year in 1986 to $50 million in 1989, and included sponsoring a variety of international mercenaries to aid UNITA's efforts.[22] In Afghanistan, US support for the Afghan Mujahedeen's efforts to repel Soviet military presence in the country was an enormous CIA operation. By the late 1980s, the US was providing an army of jihadist insurgents (many of whom were Afghan large landowners fighting against redistribution of land and many others of whom were foreign Islamists who had come from various Gulf counties) with an estimated $700 million annually in military assistance, costing the US over $3–4 billion over the entire duration of the program.[23] As in previous para-institutional engagements, private companies were often contracted to provide logistical and training support to these paramilitary armies.

In counterinsurgency settings, the US provided military aid, assistance, and training to allied regimes. These states, in turn, delegated force to non-state armed groups, sometimes under US direction and training. For example, US commitments to El Salvador's counterinsurgency program reached unprecedented levels with Reagan promising around $50 million in emergency military aid in 1981. Alongside training and arming the nation's official military, US advisors invigorated paramilitary programs by replacing ORDEN (a paramilitary program dating to the 1960s) with a civil defense force program. Numbers of local civilian irregular forces swelled to an estimated 30,000 fighters.[24] More indirectly (i.e. with limited direct US participation), US-supported counterinsurgency in Guatemala exploded in the 1980s with the recruitment of a 300,000-member civilian defense force against the insurgents, effectively forcing many of the country's inhabitants into participating in the war.[25] Similarly, in the Philippines, US military assistance and training to the country increased as the Filipino government turned to vigilante paramilitary forces. While the US had only limited or no direct role in supporting para-institutional forces, vigilante anti-insurgent squads and self-defense forces constituted a considerable force in the Philippines' US-backed counterinsurgency and by "the end of 1987, over seventy vigilante groups were reported to have formed," consistent with US training.[26] Similar structures appeared in countries such as Turkey (Turkish village guards) in the mid-1980s, amidst increased US military assistance to the country.[27]

The Reagan administration aided in the expansion of these practices in US foreign policy by advocating for them publicly and inculcating a "freedom fighter" discourse. As just one example, in contrast to the Bay of Pigs invasion which was a wholly covert affair twenty years prior, support for paramilitary fighters in the 1980s was announced in numerous official statements. In a series of speeches to the public, Congressional messages, and other communications, Reagan advocated on behalf of "freedom fighters" resisting "communist" and "socialist" forms of "tyranny," claiming that "support for freedom fighters is self-defense."[28] Reagan famously announced to the nation, in his seventh State of the Union Address, for example, that he would be seeking to extend further support to the Contra "freedom fighters" in Nicaragua. "So, too, in Afghanistan," he stated, "the freedom fighters are the key to peace." This was likened to "a swelling freedom tide across the world," connecting US support for these groups and others in Cambodia and Angola.[29] At one point, he famously went as far as to compare the Contra "freedom fighters" to the founding fathers of the United States. Somewhat paradoxically, while the implementation of support programs remained "off the books" (such as in the Iran-Contra case), this public advocacy and propaganda in support of the "freedom fighters" largely "left 'plausible deniability' in tatters" and stood "the very definition of covert action on its head."[30]

Bolstering Para-Institutional Complexes:
Right-Wing Movements and Organizations

With the anti-communist agenda set, the higher echelons of US government and the CIA conspired to collaborate with a wide variety of domestic
(i.e. within the US) and international right-wing or anti-communist
movements to further the goals of the Reagan Doctrine and expand US
paramilitary capability. The idea was to help direct public-private relationships towards US foreign policy objectives. US Secretary of the Army
John O. Marsh articulated in a US military report and conference on US
"special operations" in 1984 that "We must find a way to incorporate into
a grand strategy the total resources of our society, so as to address those
needs essential to our security beyond the limitations of our current
defense structure."[31] Scholar Sara Diamond concluded that "state policymakers saw it was in their interests to allow or facilitate direct foreign
intervention by non-governmental organizations."[32] The expertise and
influence of hundreds of war veterans and counterinsurgency experts,
as well as right-wing enthusiasts, were harnessed to contribute to the
cause of global counter-revolution. Washington coordinated these
efforts through a variety of meetings between government officials and
"private" organization leaders.[33] By the time of Reagan's second term, the
"privatization" of central tasks in this manner and the exploiting of the
influence of right-wing movements "from below" became a central facet
of the Reagan administration's foreign policy.

Throughout the 1980s, a myriad of non-governmental movements,
organizations, and businesses contributed to the growth of
para-institutional complexes in US coercive statecraft. Groups such as
the World Anti-Communist League (WACL) held various connections
in and out of US government (for instance, former CIA agent John
Singlaub) and collectively constituted a non-state means through which
Washington orchestrated a global counter-revolutionary push.[34] Groups
such as the American Security Council and WACL also held direct connections with leaders of parallel military formations (some likened to
"death squads") around the world.[35] Other organizations, such as the
Concerned Women for America, *Soldier of Fortune* magazine, Citizens
for America, Committee on the Present Danger, among others lobbied for
increased government aid to disparate paramilitary "freedom fighters",
mercenaries, and private military companies in Southern Africa, the
Philippines, Nicaragua, Afghanistan, Angola, and others, and directly

provided grassroots funding themselves to such groups.[36] International donors from governments such as Saudi Arabia and Israel, among others, and international organizations and companies also contributed. Many scholars have already uncovered and traced these developments with, for example, Diamond referring to a coalition of "right-wing power"[37] and Marshall, Scott, and Hunter describing broader "shadow networks."[38] Bodenheimer and Gould traced the emergence of a "global rollback network in which it is difficult to distinguish what is governmental activity, what is private, and what is public-private meld."[39]

Such movements and organizations played a variety of roles towards the projection of US military force through non-state or para-extended means. First, many movements and institutions aided on an ideological level by promoting "freedom fighter" propaganda.[40] Policymakers in Washington reached out to right-wing think tanks, publications, and other activist groups to "invigorate international media programs" in favor of US support to various anti-revolutionary forces around the world.[41] Similarly, organizations such as *Soldier of Fortune* magazine, Civilian Military Assistance, and the Air Commando Association produced materials depicting anti-communist insurgencies and terrorist groups, such as the Afghan Mujahedeen as "freedom fighters." This "freedom fighter" discourse aided in the justification and normalization of US para-institutional complexes.[42]

Secondly, many organizations lobbied Congress for increased paramilitarized intervention abroad and for limiting restrictions on US military aid to US-sponsored para-institutional forces. Think tanks such as the Committee of Santa Fe, for instance, published reports advocating US support to para-institutional forces abroad, including mercenaries.[43] The Heritage Foundation and the Free Congress Foundation, for example, were prominent lobby groups that went to considerable efforts to promote Congressional approval and authorization of US interventions.[44]

Perhaps most importantly, this wider privatized support network distributed materials and financial support directly to anti-communist local para-institutional forces around the world. They also sometimes participated in hostilities abroad as international mercenaries or "private" military forces.[45] Diamond, for example, details several categories of fundraisers, including US mercenary organizations, Christian Right groups, and conservative political movements, that raised funds and materials in support for anti-communist para-institutional forces. Many groups were active in supply networks to paramilitary groups abroad such as UNITA in Angola, RENAMO in Mozambique, the Contras in Nicaragua, and Filipino counterinsurgent paramilitaries, among others.[46] These supply networks were sometimes substantial. Estimates of the amount of money donated to the Contras from "private" sources run to over $25 million, with money and materials sometimes directly donated to armed groups.[47]

Proprietaries to PMCs:
The Dawn of the Modern Private Military Industry

The global counter-revolutionary push during the 1980s also expanded and consolidated the use of contracting, and the development of a modern private military industry. Within the contexts of, among others, the rise of neoliberalism and the mantra of privatization, and the perceptions of successful contracting in Vietnam, the 1980s saw the dawn of the modern PMC industry in the US. The US, due to the pressures on it in upholding its managerial role in the international system, was at the forefront of making this possible. MNCs also increasingly contracted PMCs for security of their assets and infrastructure. Oil companies, for instance, rather than relying on state armed forces, increasingly turned to PMCs for the protection of their operations. PMCs and private security forces began to operate essentially as "investment enablers" in many unstable regions, often liaising with and working towards stabilization goals in coordination with state counterinsurgent forces. Edward Herman and Gerry O'Sullivan, for example, detailed the myriad close connections between the US government and its military agencies, a burgeoning private security sector, and MNCs. They outlined how private security and military companies became a mainstay of broader US stabilization strategies and in MNCs' security repertoires.[48] State officials, MNCs and PMC corporations would increasingly interact in high-profile private security conferences, through public-private partnerships, contracting and think tank reports and other forums. During the 1980s, there developed a significant symbiosis between the "public" and "private" realms towards stabilization goals.

In the prosecution of the Reagan Doctrine, US planners relied to a greater extent than ever before on contracting to private military groups during the 1980s. US contracts with PMCs, including former CIA proprietary companies such as SAT and Summit Aviation, increased in volume as well as in the breadth of services provided throughout the 1980s. The CIA continued to contract such companies for covert missions performing logistics and limited combat roles in hot spots such as Nicaragua and El Salvador. They were also increasingly hired for a wider variety of activities by the US Department of Defense, primarily logistics, such as for the Pentagon's Military Airlift Command.[49] The *New York Times* reported in 1987 that SAT, after being "privatized", "now has 25 planes and 540 employees, almost double that of a year ago, and its earnings—$907,000 in the third quarter of 1986, and $1.24 million in the second quarter—have risen sharply." Fletcher Prouty, a retired Air Force Colonel summarized these developments in the 1980s: the "agency [CIA] uses fewer wholly subsidiaries and more private contractors, but the range of activities is little changed, and the volume of business [of the private contractors] could be 10 times higher than in our day."[50]

As part of this growth, the 1980s saw the rise of modern PMCs. Following mounting scrutiny of CIA activities in the 1970s intelligence reviews (such as the Church Committee) and in accordance with a "privatization logic," most of the covert CIA-run semi-private proprietaries operational during the previous decades were completely sold off, often to former military or CIA personnel with continued connections to Washington. These enterprises remained contractors operating as independent businesses. The CIA sold its shares in its semi-private proprietaries between the mid-1970s and early 1980s.[51] This included a variety of airline and logistical service companies financially linked to the CIA around the world.[52] According to one CIA document, the CIA "anticipated that Air America, Inc., Air Asia Company Limited, Civil Air Transport Company Limited, Air America Limited and Thai-Pacific Services Company Limited could be sold on a going concern basis and continue to operate in the private sector."[53] In some cases, such companies were totally liquidated and redeveloped into new ones; in other cases, they continued contracting to the US government. For example, Southern Air Transport, an airline active in a variety of contracted services for US military missions, was sold off in 1973 to Stanley Williams, a former SAT manager for the CIA, and continued to contract its services to the CIA and a variety of other buyers across the world.[54]

The CIA divested itself of shares in Summit Aviation and the company continues to operate today in Delaware as a private air-contractor. Its website lists the Department of Defense and the CIA as employers for "special missions."[55] SAT continued as an air-contractor into the 1990s.[56]

The history of DynCorp (still an active and lucrative business) is also instructive. The company has its origins as a private airline contractor (like CAT or AA) in 1946 (then under the titles Land-Air Inc. and California Eastern Airways [CEA]). After serving as an airlift capability in the Korean war, CEA was awarded the first Contract Field Team (a US Air Force program to solicit private support for technical aviation services) contract in 1951 for "depot-level repair to U.S. military aircraft and weapons systems worldwide," and has been awarded similar logistical contracts ever since.[57] In 1951, revenue from such contracts hovered around $6 million.[58] The company, diversifying beyond the defense industry, became DynCorp in 1987 and has subsequently been contracted for a variety of military missions.[59] For instance, by the late 1990s, DynCorp had an integral role in providing counterinsurgent military assistance to Colombia and also trained forces in El Salvador and Haiti.[60]

In addition, after successful uses of contractors in Vietnam, new businesses were also created. Private companies offering militarized services were also contracted to do a wider variety of activities than ever before, and by an increasing number of conventional US departments and agencies. Rather than solely a tool of covert action (although this was still a primary arena in which contracting occurred), Washington's use of private military companies and mercenaries became further entrenched in official US foreign policy channels.

US planners considered contracting and the use of para-institutional forces during Vietnam a success and sought to apply these lessons elsewhere. For example, contracting a variety of US military services to private companies was considered an efficient and cost-effective way to overcome logistical issues. An official US Army post-Vietnam report, for example, concluded that "The successful techniques and procedures developed by U.S. Army Procurement Agency, Vietnam in providing these procurement services, in the combat zone, will be the basis for contract logistical support in future conflicts."[61] It outlined the successful use of contractors and the ways in which the experiences with contracting during the Vietnam War provided the basis for the post-Vietnam development of institutional frameworks to administer military contracting. Scholars such as Dickinson confirmed that "official military reports after the war make the case for continued and increased use of contractors to provide logistical support on the battlefield."[62] In his studies on the history of contracting, Carafano also found that "The Pentagon had actually considered contracting in combat during Vietnam a big success."[63] US planners saw the use of contractors for covert operations to have significant advantages that could be applied more generally and this contributed to the formation of new PMCs.[64]

Many prominent private military companies were created in the 1980s, before the privatization boom of the 1990s.[65] For example, "eight highly skilled and experienced military leaders" founded MPRI in 1987.[66] Company operations however, expanded significantly after the end of the Cold War and have been involved in training numerous foreign militaries. Former US ambassador to the UN, Jeane Kirkpatrick was part-owner of Operations and Policy Research, Inc., a company that advised the government on counterinsurgency and risk analysis in low-intensity conflict environments.[67] Two former US military officers founded Airscan in 1989, a company which became integral to US drug-crop eradication programs in Colombia during the 1990s and 2000s.[68] The Betac Corporation has also been a key player since the early 1980s. According to Shorrock, Betac is "a consulting firm composed of former intelligence and communications specialists from the Pentagon." It "was one of the largest government contractors of its day and, with TRW and Lockheed itself, dominated the intelligence contracting industry from the mid-1980s until the late 1990s."[69] Eagle Aviation Services Technologies, for instance, set up by Richard Gadd, a retired Air Force officer in 1982, was contracted for missions in Nicaragua and was later contracted for a variety of other missions such as in Colombia's drug war during the 1990s and early 2000s, and continues to provide services to the US government.[70] A former Air America chief pilot founded a company called Aero Contractors in 1979, which subsequently provided a wide array of contracted services to US agencies and departments.[71] Aero Contractors was later linked to the US's program of "extraordinary rendition" during the "War on Terror."[72] The US became the leading home or base for PMCs (alongside the UK and South Africa) and the US government emerged as the primary consumer of their services, dedicating increasing portions of its military budgets towards PMCs.

US military planners also established many of the institutions and the administrative infrastructure for contracting to institutionalize and regulate outsourcing of military logistics and other roles, which also helped to consolidate privatization into US military affairs. Based on the temporary offices of the United States Army Procurement Agency Vietnam, set up to contract logistical requirements during the Vietnam

War, the US Army created the Logistics Civil Augmentation Program (LOGCAP) in 1985. LOGCAP was an administrative body that "set out the concepts, responsibilities, policies, and procedures for using civilian contractors to replace soldiers and recruiting local labor during wartime."[73] LOGCAP was a forerunner to the procedures for outsourcing that are currently used. Similarly, the Diplomatic Security and Antiterrorism Act in 1986 allowed the Department of State to contract private security for diplomatic missions.[74] Rather than covert arrangements to obscure the US's role in intervention, the institutionalization of outsourcing practices reflected a stabilized set of relations between the US state and various private for-profit military enterprises. These developments paved the way for the rise of the private military industry and its subsequent significant growth in the 1990s. It was under Reagan's reign that the contemporary private military industry in the US was born.

The rise of neoliberal ideologies under Reagan's tenure also contributed to this increased para-institutionalization in the 1980s and 1990s through the logics of "privatization." Neoliberal capitalism emphasized limited government direction of an economy in favor of the market efficiency of the private sector. This entailed policies of liberalization, fiscal austerity, deregulation, and privatization. These sets of "free-market" theories are often associated with the work of Frederick Hayek and Milton Friedman. Although neoliberal ideologies and their application in political and economic restructuring in the US pre-dated the Reagan administration, neoliberalism accelerated under Reagan's tutelage, along with a penchant for privatization.[75] These reforms represented a broad shift in economic thinking, as well as political and social policies and practices across sectors. However, these assumptions about market efficiency and corresponding approaches to cost-effectiveness aided in the expansion of para-institutional complexes both directly and indirectly.

More directly, the Reagan administration implemented policies conducive to privatization across different sectors, including, very gradually, in US military activities. The Reagan administration sought to systematically implement Circular OMB-76, with amendments (1983) designed to gauge government competitiveness with the private sector with the aim of enhancing cost-effectiveness.[76] The number of contracts awarded to businesses to conduct government activities more generally (i.e. not necessarily military) grew by 10 percent between 1983 and 1985, and significantly more in Reagan's second term.[77] The logic of OMB-76 was increasingly applied to areas of the military, primarily logistics throughout the 1980s and 1990s, with the aim of increasing efficiency and cost-effectiveness driving contracting. Krahmann notes, for example, that Reagan's neoliberal logics of privatization were applied to the military through the Competition in Contracting Act in 1984 and two further similar Acts in 1986 which had the effect of "greater reliance on the resources and expertise of private defense contractors for the development, production and provision of military equipment and services."[78] This led to the development of LOGCAP in 1985, as briefly discussed above.[79] Under LOGCAP, the DoD awarded the first major long-term umbrella contract in 1987 to a private firm to tend to non-core US military capacities, such as logistics.[80]

More indirectly, the privatization push helped to entrench perceptions of the efficiency and cost-effectiveness of non-state actors in coercive roles. Ideas of privatization fueled the development of a PMC industry in the 1980s as people in and out of government gradually began to see private business as legitimate harbingers of official state action.[81] For example, the retired US military generals that founded MPRI (a prominent US-based PMC) in 1987 not only looked to capitalize on the increasing application of OMB-76, but also understood how the appeals of privatization would allow them to sell their expertise back to the US government.[82] In addition to this, after the withdrawal from Vietnam, the US ended military conscription and established an all-volunteer force in 1973. This opened up the recruitment of military personnel, the organization of logistics, capital (equipment), indeed, the entire military apparatus, to competition with the market.[83] It also had a cultural effect of transforming the understanding of military service from a logic based on citizenship to the logic of the market.[84] While these effects were not immediate, the creation of a volunteer force helped to create the conditions conducive to the rise of the private military industry in the US in the 1990s. The privatization push also provided the space into which a nascent PMC and private security industry grew.

The Para-Institutional War on Nicaragua

Washington's war against the Nicaraguan Sandinista government serves as an example of some of these developments. The prosecution of US statecraft in Nicaragua was delegated to para-institutional forces to unprecedented levels and in new ways. Alongside the Contras, as the Nicaraguan right-wing militants were known, the US coordinated mercenary outfits, air-contractors, and PMCs for the destabilization of the Sandinista regime. PMCs were increasingly involved in various aspects of US unconventional warfare and featured prominently at the forefront of battle and for paramilitary logistical requirements. While the unconventional war was fought on the front lines by a wider variety of paramilitary agents, Washington sought new privatized means of funding and providing assistance to these groups. Washington leveraged the power of a wider set of private actors, including networks of civilians, NGOs, and other private companies and organizations, to fund and facilitate its low-intensity conflict against Nicaragua. As Marshall, Scott, and Hunter described it, "President Reagan's secret weapon is 'contracting out' such normal government functions as funding and executing policy to the 'private' sector while keeping policy making itself in the hands of the state."[85] Washington's para-institutional conflict was coordinated to coincide with and reinforce parallel political and economic pressures against Nicaragua.[86]

Following the overthrow of the US-supported Somoza regime in 1979, Sandinista revolutionary reforms guided by the "logic of the majority" promised to tackle poverty and exploitation, creating a system of political and economic pluralism, much to the consternation of the country's elites who saw such maneuvers as a "betrayal."[87] Based on a platform of non-alignment (i.e. neither aligning itself with the US or the USSR), the political and economic transformation of Nicaragua threatened to lead Nicaragua away from the US sphere of influence. As Kolko states:

> Nicaragua, like Cuba before it, was of profound significance in the United States' relationship to the hemisphere, and both confirmed that it had irrevocably lost its ability to control the main political developments that grew irresistibly out of the economic policies and social forces it supported. Nor could it stem the political consequences of United States' endorsed structural changes or define alternatives to them, for these impinged on its own basic economic needs and interests as well as those of the classes with which it was aligned.[88]

Moreover, the appeal of the Sandinista revolutionary platform, in the eyes of US policymakers, threatened to spill over and spread to other areas. Refracted through the lens of the Cold War logic, this signaled the potential for Soviet influence in the region and the possibility of a contagion of revolutionary social change.

Around 1978–80, the Carter administration initiated covert support to right-wing political parties, trade unions, and favorable media outlets. However, it was during the Reagan administration that destabilization measures snowballed. Upon Reagan's inauguration, the orchestration of an insurgent war began with the implementation of the National Security Decision Directive on November 23, 1981, allowing the CIA to train and direct ex-Nicaraguan National Guardsmen and Somocistas. The Contras grew from 500 members in 1981 to around 15,000 in 1984, with many receiving salaries from the CIA.[89] Reagan authorized an initial $19.5 million towards those ends and another $50 million to Argentina to train some of the Contra fighters.[90] The total sum of covert (but official US government) military aid throughout the conflict to the Contras—including funds presented as "humanitarian aid," other portions allocated through contingency appropriations, and those financed clandestinely—is difficult to estimate. However, estimates place it in the tens of millions annually.[91]

The CIA trained the Contras in insurgency, "sabotage" techniques, demolition, and explosives in various covert locations, including Argentina, Honduras, and the US. The CIA manual *Psychological Operations in Guerrilla Warfare*, which was used to train the Contras, advocated "armed propaganda," killing of judges and other civilian officials, amongst other activities aimed at fostering a popular insurgency against the Sandinista government. Following this training, Contra fighters conducted bombing campaigns in Managua, the Nicaraguan capital and the main international airport, from bases in neighboring Honduras and Costa Rica.[92] By the end of 1985, the Contras had reportedly killed 3,652 civilians and kidnapped 5,232, gaining a reputation for their brutality, which included mutilating many of their victims.[93]

While the Contras waged their deadly insurgency, in 1983 and early 1984, the CIA employed "Unilaterally Controlled Latino Assets" (UCLAs): trained mercenary commando teams from Honduras, Guatemala, and other Latin American countries. The UCLAs' mission was to support the Contras, disrupt trade, terrorize the Nicaraguan people, and conduct bombing raids and "sabotage" operations against key installations and areas of strategic and economic importance. For instance, the UCLAs, operating from CIA boats, were responsible for exploding an oil facility in the port of Corinto (injuring 100 and temporarily displacing 25,000). Additionally, the UCLAs mined Nicaragua's main commercial harbor, with the ultimate aim of crippling the

Nicaraguan economy.[94] One Honduran UCLA later remarked that their mission was "to sabotage ports, refineries, boats and bridges and make it appear that the Contras had done it," in order to create the impression the Contras were significantly stronger than they were.[95] The UCLAs were an additional mercenary asset at CIA disposal to further debilitate the Sandinista regime while simultaneously preserving a perception of US restraint. Nicaragua later filed suit against the US in an international court for the mining of its waters.

Mercenary organizations from the US not only provided training and assistance to Contra groups, but also conducted their own operations.[96] The American group called Civilian Military Assistance (CMA), for instance, comprised numerous Vietnam veterans and was established in 1983 to aid the "counter-revolutionary" agenda. Their participation in the Nicaraguan war became known when three former US military personnel were killed after their helicopter was shot down after conducting an attack on the Nicaraguan town of Santa Clara in 1984.[97] In addition, a Washington-based group called GeoMilTech, a small military consulting firm, shipped $5 million worth of arms to the Contras.[98] GeoMilTech's executive board at the time comprised of numerous influential former policymakers from Washington, including General John Singlaub (also head of the World Anti-Communist League), John Carbaugh, and Robert Schweitzer (a former US Army general).[99] Similarly, in 1985, five mercenaries (from the US, the UK, and France) were arrested in a Contra training camp in Costa Rica.[100] These international parallel military organizations such as CMA, Soldier of Fortune, and Air Commando Association were mobilizing veterans and ex-US military personnel for various missions from military engagements, piloting planes and helicopters, to training and running supplies to the Contras, as well as fundraising.

PMCs contributed an additional layer of para-institutional fighters, which much like the UCLAs and paramilitary organizations, were employed to conduct special operations and training exercises. In 1984, the CIA approached David Walker, a former British SAS (Special Air Service) commander, to contract military operations out to his PMC Keenie Meenie Services (KMS). According to North's later testimony to the Congressional committees investigating the Iran-Contra affair, Walker offered to conduct "sabotage operations for the resistance," inside the capital Managua and elsewhere, including a $50,000 contract to destroy Nicaraguan Army ammunition depots and plans to demobilize helicopters at Managua's main airport.[101] Walker and his KMS were authorized to conduct other military and "sabotage" operations in Nicaragua, which according to Oliver North in later testimony were intended to foster the "perception that the Nicaraguan resistance could operate anywhere that it so desired."[102] Walker and KMS recruited an estimated fifty mercenary members mainly from the US and the UK to complement the Contra forces.[103] Another un-named PMC was allegedly hired to conduct "so-called policies of intimidation."[104] In addition, the CIA planned for Walker's introduction to Calero (a Contra group leader) in order to arrange special operations and insurgency training programs for Contra troops. While Calero was to contract KMS directly (thereby not directly implicating the CIA), efforts were made to "defray the cost of Walker's operations from other than Calero's limited assets."[105]

With media coverage of these actions leading to public outcry, Congress sought to limit US assistance to the Contras. The Boland Amendments (1982 and 1984)[106] prohibited US military assistance to the Contras. This meant devising new ways to circumvent and overcome these restrictions if aiding the Contras was to continue. This dynamic had a particularly strong effect on the provision of assistance to the Contra insurgent forces. In lieu of absence of authorized assistance, funding for the Contras was secured through various "private" sources and third-country donations, such as from Saudi Arabia and Taiwan, with estimates of totals raised running up to $100 million between 1983 and 1985. Such initiatives saw campaigns for "Christmas bags" donated to Contra guerrillas along with food, clothing, and medical supplies. Funds and military equipment were often channeled through the CIA from private donors or companies. For instance, a disused fleet of twenty Cessna aircraft were procured from the New York National Guard and channeled via Summit Aviation, Inc. (a former CIA proprietary) under contract to the Contras.[107] Washington continued covertly via "private" means what it could not accomplish officially. To coordinate, supply, and facilitate Contra efforts, CIA staff Oliver North and William Casey established what they called the "Enterprise"—a network of private organizations and off-shore bank accounts. The Enterprise was created explicitly to bypass Congressional scrutiny and to provide the US with the plausible deniability it needed in order to be able to sustain its paramilitary war.[108] The Enterprise remained the principal channel through which the US state financed and coordinated the Contra forces until 1987, when Congress voted for the resumption of aid, approving $100 million to the Contras.

The CIA also employed PMCs to perform logistical requirements in providing military assistance to the Contras. Enterprise coordinators North and Richard Secord arranged for retired Air Force Lt. Col. Richard Gadd to coordinate Contra resupply efforts using various private companies. Gadd contracted SAT which he interlinked with his own set-up of private airlines such as Eagle Aviation Services and Technology (EAST), and American National Management Corporation (ANMC), procuring airplanes for the delivery of "lethal" assistance in contravention of Congressional bans. All these private companies also secured contracts with the US government for other operations such as logistical flights for the Grenada invasion and for Special Force transportation for "low-visibility operations" and training exercises in the Caribbean.[109] The transfer of materials and weapons to the Contras was conducted by private airline companies—SAT (receiving around $2 million), Corporate Air Services (contract worth $437,000), and EAST (contracted for just over $600,000)—with much of the money towards the resupply operations laundered through private companies with connections to Gadd.[110]

Pilots hired to fly resupply missions and other combat roles in private airlines in Nicaragua were either retired Air Force personnel or were "sheep-dipped" and employed as a civilian. But later, at North's request, the Enterprise paid David Walker (of KMS) $110,000 for two pilots to fly supply missions to the Contras in order to avoid using US military personnel.[111] The issue, according to Secord in later testimony, was an "appearance problem. If we were to have one or more of these people captured, as it ultimately occurred, it becomes a real problem when it's American citizens."[112] Here, Secord makes reference to the capture of former US marine Eugene Hasenfus in 1986 by the Nicaraguan authorities after his plane was shot down. The other crew members, who were in possession of SAT identification cards, died, while Hasenfus's capture went on to expose the Iran-Contra scandal. It was later revealed, although the State Department issued a statement denying US involvement, that knowledge of the event and attempts to scramble a rescue operation went up to the highest echelons of government. The Enterprise had also hired Col. Robert Dutton, retired Air Force officer, to replace Gadd to coordinate resupply efforts on the Southern Front.[113]

The outsourcing of US coercive apparatus in this case demonstrates the fluidity by which ex-officials could continue to use their influence, knowledge, and training for personal gain, while serving as a "private" extension to public objectives. One of the most influential personalities in this regard was retired Gen. John Singlaub, a CIA veteran of paramilitary operations and a counterinsurgency expert, who not only served as an independent Pentagon advisor but also headed the WACL during much of the 1980s, WACL being another organization which helped to rally domestic support for Reagan's proxy wars. Singlaub was also affiliated to groups such as the National Defense Council and Air Commandos Association, which collectively coordinated their actions towards Nicaragua through the NSC and the Enterprise system. This amounted to a specific approach to outsourcing, tapping into the reservoirs of expertise of ex-servicemen. As Marshall, Scott, and Hunter explain, "Unlike typical commercial examples of the practice," this "administration has contracted to agents who are themselves total creatures of government—in particular, of government intelligence agencies. In their 'private' capacities, however, these agents nonetheless fall largely outside Congressional purview."[114]

These patterns and relationships helped fuel the growth of a PMC industry in the US in the early 1990s. While the US contracted PMCs previously, it was during the 1980s that the PMC began to take its present corporatist form. Retired servicemen created their own companies or were hired by various branches of the US state for their expertise and experience. These relationships represent precursors to the growth of the PMC industry in the early 1990s, whereupon ex-servicemen could sell their knowledge and skills as part of an official business.

The nature and extent of the outsourcing of the Contra war indicates an increasing reliance of the US state on parallel military groups to execute coercive interventions. This forms part of a long-term pattern in which the US has been unable or unwilling to intervene directly to assert its interests and influence. Ambitious plans for the "horizontal escalation" of low-intensity conflict across the global South, or as one military expert put it, the "strategy of worldwide war," required leveraging support from other sources to meet the demands of the Reagan administration.[115] Outsourcing was part of a deeper process that involved a nexus between state officials and a collection of para-institutional forces, a series of public-private partnerships. The Reagan Doctrine, and the ancillary low-intensity conflict strategy ultimately giving way to a series of outsourced military projects, were part and parcel of the US's grand strategy.

Conclusion

During the 1980s, there was a gradual escalation of a nexus between the US and a variety of para-institutional forces. There was an evolution in the use of global/local para-institutional forces in US foreign policy from what emerged as an instrument of covert operations in the late 1940s and early 1950s, to expanded logistical and combat functions in Vietnam and beyond, and in the institutionalization of such practices in the 1980s. Practices of outsourcing were largely consolidated in US counterinsurgency and unconventional warfare modes of imperial statecraft. The classic covert practices of supporting local militias, paramilitary groups or insurgents, complete with the help of semi-private airlines and PMCs for their resupply and military support, slowly gained prominence as an acceptable alternative means of extending US coercive reach whilst preserving an image of US restraint.

Towards the end of the Cold War, US planners predicted that internal conflicts in the global South would continue to be the principal threats to US national interests and security. Yet rather than a continuation of a geopolitical battle with an arch super-power, they were cognizant of the detrimental effects of destabilizing forces within strategically important countries. For example, the Commission on Integrated Long-Term Strategy predicted in 1988 that insurgencies and other Third World conflicts would continue to

> ... have an adverse cumulative effect on US access to critical regions, on American credibility among allies and friends, and on American self-confidence. If this cumulative effect cannot be checked or reversed in the future, it will gradually undermine America's ability to defend its interests in the most vital regions ... .[116]

As we shall see, the practices outlined above informed the consolidation of US paramilitary capabilities well into the post-Cold War era and into the "War on Terror" as US planners continued to consider operating through "surrogates" and private businesses necessary for the expansion of a US-led liberalized global order.

5

Continuity After the Cold War and the Consolidation of Para-Institutional Complexes

The demise of the USSR and the wane of the ideological challenge of communism at the end of the Cold War ushered in significant transformations in the international system. With the dissolution of the USSR, the superpower tensions that characterized the Cold War dissolved and the US emerged the unquestioned leader of an emerging "new world order." Meanwhile, obstacles to capitalist powers and to the fluid functioning of global capitalism in the form of communist and socialist ideological political and social programs slowly dissipated. The stabilization of capitalist socio-economic relations and the opening up of economies in the South were challenged less frequently by communist- and socialist-inspired political movements and revolutionary programs. Intellectuals and policymakers hailed the "end of history," of the end of an international ideological struggle, and the US and international organization promoted neoliberal forms of capitalism in the form of a "Washington Consensus" as a standard model for development, and political and economic organization.

The implications of the end of the Cold War were that the US, in conjunction with its Western allied states, was relatively free to assert its managerial role in the international system and pursue the expansion and maintenance of a US-led liberalized global order. During the 1990s, US policymakers enhanced US-sponsored globalization through international organizations such as the World Bank and the International Monetary Fund (IMF), along with the creation of the World Trade Organization (WTO) in 1995 towards an increasingly global capitalist system. However, in the context of unchecked US power projection and as the leading hegemonic state in the international system, the US continued to play a lead role in the opening up of economies in the global South and in the stabilization of capitalist socio-economic relations and their corresponding political and economic frameworks. Where instabilities more generally existed—be it in the form of lingering communist insurgencies, anti-US and anti-capitalist subversion, nationalist rejections of neoliberal rule, civil unrest, and dissent, or civil conflicts—the US worked with key allied political and economic elite governing strata to stabilize favorable state arrangements abroad. Counterinsurgency and unconventional warfare remained the US imperial statecraft of choice.

In this context, the end of the Cold War and the international environment of the 1990s propelled the further entrenchment of para-institutional complexes in US imperial statecraft. Primarily, during this period, there was significant growth of the PMC industry. The US, at the forefront of broader shifts in the global economy and the rise of neoliberalism, led developments in the PMC industry more globally. US planners contracted out a wider variety of tasks to PMCs than ever before. The post-Cold War era also signaled the PMCs "officialization" and the start of their global spread. The US had led developments in the privatized military industry, but now other countries around the world had begun to accept them as well. US departments and agencies increasingly contracted PMCs for a variety of more central roles in US statecraft, including more logistical and specialist technologies and equipment, and maintenance, as well as supplementing US worldwide training programs and operating in more front-line intelligence and military missions.

In addition, MNCs more frequently contracted PMCs for private security and protected investment environments. Conducting business abroad, particularly in unstable political climates in the global South, carried risks. MNCs increasingly demanded security services to protect their assets and operations. This constituted a global trend connected to the expansion of capital in processes uncontained or beyond US foreign policy. However, as will be argued in this chapter, these arrangements often contributed to broader US-led stabilization and counterinsurgent efforts. US advisors and host-country military forces often coordinated with MNCs for the provision of their security in counterinsurgent contexts. These produced wider complexes of para-institutional forces towards stabilization goals. Rather than petitioning or lobbying state authorities for stabilization, as MNCs often did during the early Cold War, MNCs often became agents within the broader structures and sets of para-institutional relationships that worked towards stabilizing local state arrangements conducive to US and transnational capital interests.

To make these arguments, this chapter first briefly analyzes the continuity of US imperialism in the post-Cold War. It then turns to the continued promotion of localized paramilitary forces, highlighting the persistence of the strategic rationales and sets of interests contributing to their presence in US-backed counterinsurgency and unconventional warfare. This chapter then turns to the rise of the PMC industry. It argues that while existing studies have captured the myriad factors that computed into a rise of the industry as a whole, they are insufficiently attentive to the ways in which US-based PMC activities were tied to US imperial strategies. It then provides a variety of examples of the increased prevalence of PMCs in US foreign policy and the ways in which they have helped bolster US power projection. Finally, it examines the US-supported counterinsurgency campaign in Colombia as an example of these trends. During the 1990s and early 2000s, Colombia was one of the largest recipients of US counterinsurgency assistance, particularly with the passage of Plan Colombia in 2000 and subsequent military aid packages, and therefore represents a likely case in which these post-Cold War dynamics would unfold.

Imperialism and "Stability" in the Post-Cold War Era: A New World Order?

The easing of East–West tensions with the dissolution of the Soviet Union and the fall of communism did not fundamentally alter North–South relations. If anything, the decline of communism as a viable political-economic alternative and the absence of an oppositional superpower meant the US was less constrained to enforce an expanding liberalized global order.[1] The US emerged as the unquestioned leader of liberal democratic capitalism, strengthening the US's position in the political and military enforcement of the emerging order in an ideological "end of history."[2] Simultaneously, developments in global capitalism provided a further basis for the increasing trans-nationalization of capital.[3] Accelerated neoliberal policies packaged via a "Washington consensus" became a standard developmental paradigm for international institutions, contributing to the further liberalization of national markets around the world. Within this context, the US continued to serve as the guarantor of last resort for stable investment climes in the global South working towards US and transnational capital interests. Thus, the logics underpinning much of US foreign policy, particularly towards the

South, remained largely intact in the post-Cold War period. As many scholars and observers have already argued, US Open Door statecraft continued to concentrate on creating and maintaining conditions abroad conducive to mutually reinforcing transnational political and economic interests.[4]

Consistent with this, where direct contestation of the prevailing order emerged within countries in the South, US planners relied on time-tested counterinsurgency and unconventional warfare politico-military strategies to conserve favorable state arrangements abroad. Statements contained in numerous national security documents help to point to these continuities in the US foreign policy after the Cold War. The then Secretary of Defense Dick Cheney predicted in his *Annual Report to the President and Congress* in 1990 that "low-intensity conflict will remain, as it has since 1945, the most likely form of violence threatening US interests." Internal conflict, he further suggested, which "largely results from instability in the Third World," presents "a real and immediate danger to democracies, and threatens relationships and alliances that are vital to the coalition defense and open economies of the United States and its allies and friends." In defending these interests of both the US and transnational elites, Cheney advocated a politico-military response which hinged on "winning popular support" in host countries "rather than merely capturing and controlling territory." Therefore Cheney viewed it vital to continue to employ strategies "that rely more heavily on mobile, highly ready, well-equipped forces and solid-power projection capabilities."[5] Special Forces, he noted, were best positioned to work with, through, or by local forces, coupled with military assistance to recipient country's official military armed forces to strengthen their defenses against internal centrifugal forces. Similarly, General A.M. Gray wrote in a 1990 policy report:

> If we are to have stability in these regions, maintain access to their resources, protect our citizens abroad, defend our vital installations, and deter conflict, we must maintain within our active force structure a credible military power projection capability with the flexibility to respond to conflict across the spectrum of violence throughout the globe.[6]

In the absence of an overarching ideological contestation to the US-led liberalized global order, manifest in communism during the Cold War, "instability itself" was increasingly recognized as a major threat to the "new world order" articulated by President Bush.[7] The 1991 National Security Strategy, for example, stated that in this bid "to build a new world order in the aftermath of the Cold War, we will likely discover that the enemy we face is less an expansionist communism *than it is instability itself.*"[8] In this way, US planners understood that "national security and economic strength are indivisible," whereupon the US's defense policies seek primarily "to promote a strong, prosperous and competitive U.S. economy; ensure access to foreign markets, energy, mineral resources, the oceans and space; promote an open and expanding international economic system."[9] Such iterations of the need for the promotion of "stability" are found in national security statements throughout the 1990s.[10]

Military planners continued to view counterinsurgency and unconventional warfare, often euphemistically referred to as "low-intensity conflict" or "military operations other than war", as important tools in the pursuit of reinforcing the "new world order."[11] According to Grosscup, US military planners in the early 1990s envisioned that the use of terrorism arising from nationalism, irredentism, and religion (rather than communist ideology) would make "low intensity conflicts of the 1990s even more threatening than those of the Cold War."[12] As such, security assistance to the global South in the form of military-to-military ties and counterinsurgency training were sustained as the principal politico-military instruments of statecraft in the forging of the desired "stability." US military training programs expanded significantly in the post-Cold War era to an increased number of countries, with the International Military Education and Training (IMET) program growing four-fold between 1994 and 2002.[13] Much of this training continued to have a counterinsurgent orientation geared towards protecting the prevailing domestic order and internal policing and security. Consistent with this, US interventions in the global South, either covert or overt, increased during the Clinton administration.[14] These interventions relied on outsourcing to collections of para-institutional forces. US planners directed Special Forces and the CIA to liaise with local military non-state forces as well as private military companies. Consistent with these US policy objectives, one Special Force training manual noted how "the objectives of security assistance are to support U.S. national security interests and strengthen the military capability of selected friendly and allied countries." This, it envisioned, would

... foster favorable attitudes toward the United States and its policies; encourage friends and allies to pursue national objectives compatible with U.S. foreign policy and military strategy ... Develop defensive self-reliance of other nations, thus reducing the need to commit U.S. forces in local crisis situations.[15]

Consistent with previous Cold War engagements, US military counter-insurgency training to recipient states in the global South continued to provide strategic rationales for leveraging pro-government militias and paramilitary forces to defeat counter-hegemonic armed insurgencies and non-armed social movements. One manual read: "Commanders must influence (rather than dominate) their operational environment to create favorable politico-military conditions for achieving specific national security objectives."[16] In order to do this, it stated, they must "apply military power indirectly through the military and paramilitary forces of a foreign government or other political group." Special Forces remained one of the primary intermediaries of this nexus through which the US could exert leverage indirectly. The 1997 Secretary of Defense annual report to the President detailed how Special Forces serve as "force multipliers" establishing "rapport with foreign military and paramilitary forces."[17] One US military manual advised that Special Forces were designed specifically "to advise, train, and assist indigenous military and paramilitary forces. The supported non-US forces then serve as force multipliers in the pursuit of US national security objectives with minimum US visibility, risk, and cost."[18] Other training manuals in the post-Cold War period detailed how Special Forces "advise and assist irregular HN [Host Nation] forces operating in a manner similar to the insurgents themselves, but with access to superior ... resources."[19] These paramilitary operatives may also be trained as "stay-behind cadres" according to one manual, in case of a hostile government takeover whereupon they could later serve as insurgents.[20] These entries and similar statements peppered throughout the US counterinsurgency doctrine, underscored the extent to which support was given to local "irregular" forces and paramilitary outfits at the heart of the counterinsurgency effort in order to bolster "regular" military capabilities of allied states.

Particularly instructive of the politico-military logics that underpinned the paramilitary proscription was the description of "civilian irregular defense forces" or "civilian self-defense forces" (CSDF). In a continuation of "counter-organization," US manuals promoted mobilizing civilians into "self-defense" units to gain access to local information and generate support for the local government amongst recruits and the populations in which they were based. Appendix D of *Foreign Internal Defense: Tactics, Techniques, and Procedures for Special Forces* (1994) shows how the self-defense forces paramilitary concept, similar to that applied in Vietnam, was supposed to divide and polarize societies to gain the active participation of members of the public. It stated that when the strategy is implemented, "the insurgents have no choice; they have to attack the CSDF village to provide a lesson to other villages considering CSDF." Yet the insurgents' response with violence has utility: "the psychological effectiveness of the CSDF concept starts by reversing the insurgent strategy of making the government the repressor. It forces the insurgents to cross a critical threshold—that of attacking and killing the very class of people they are supposed to be liberating," thereby severing the ties between the insurgents and local populations.[21] By denouncing insurgent violence and forming a link with anti-insurgent members of the public, it is hoped that the insurgents lose the battle for legitimacy for themselves by attacking civilians organized against them. This formed part of the expected dynamics of violence.[22] Paramilitarism, and the subcontracting of defense and combat capabilities to local civilians formed part of a broader political imperative to win the allegiance of the civil population. One way to do this was to mobilize them on your behalf.

The Para-Institutionalization of Counterinsurgency and Local Collaboration

Para-institutional coercive complexes were integral to protracted local state counterinsurgency campaigns. Sometimes US support for para-institutional dynamics during this time period was limited to intractable conflicts left over from the Cold War. In Guatemala, for example, after considerable US support for organizing local civilians into "civilian self-defense patrols" (PACs—*Patrullas de Autodefensa Civil*) throughout the Cold War, and after the integral US role in the spike in counterinsurgency violence during the 1980s, US military assistance to Guatemala was cancelled on human rights grounds in 1990.[23] Despite this ban on military assistance, however, the CIA supported the Guatemalan counterinsurgency campaign to the tune of $5–7 million annually throughout the early 1990s.[24] Meanwhile, as a number of US human rights organizations documented, the Guatemalan Armed Forces and their paramilitary PACs continued a violent, protracted counterinsurgency war against internal dissidents. A 1995 State Department report on human rights in Guatemala, for instance, recorded that an "estimated 340,000 men serve in rural civil self-defense committees ... [and have] committed numerous serious human rights violations and generally enjoyed impunity from the law."[25] These para-institutional structures were finally dismantled in 1996 with the signing of the peace accords.[26]

Comparable para-institutional relations also continued in the Philippines throughout the 1990s and were later expanded as part of the "War on Terror" with US influence. As a former US colony, the Philippines had long-term and strong military ties with the US ever since its independence in 1946, which primarily revolved around the fortification of military and police structures for internal policing and the eradication of communist-inspired forces and Islamic nationalist rebellions.[27] Much like similar conflicts in Latin America and elsewhere, the Filipino authorities both officially armed and covertly supported non-state military forces as a central component of the counterinsurgency campaign.[28] This was silently condoned by US military tacticians and military trainers, who held significant influence over the conduct of the Filipino counterinsurgency measures.[29] These para-institutional relations reached their peak in the low-intensity conflict environment of the 1980s, in which the US-supported counterinsurgency war oversaw networks of civilian militias and other vigilantes often generally referred to as "Civilian Volunteer Organizations" that spearheaded much of the state's counterinsurgency drive, accused by numerous human rights organizations of serious abuses and terrorizing local populations.[30]

In the post-Cold War period, as the communist insurrection in the Philippines waned, the US withdrew its military presence as it dismantled its military bases alongside its military assistance packages (ending in 1992). During this time, though, the Ramos administration (1992–98) continued to rely on a state-sanctioned and administered paramilitary militia called the Civilian Armed Force Geographical Unit (CAFGU), as well as other smaller para-institutional armed groups for counterinsurgency efforts against remaining pockets of communist forces. Amnesty International reported President Fidel Ramos stating the CAFGU were required in areas where the NPA [New People's Army—an armed communist insurgent group] were still active to preserve "stability" in the country.[31] One Human Rights Watch report recorded that since this time, "Successive Philippine administrations have publicly committed to disbanding CAFGUs, vigilante groups, and so-called private armies from time to time, but efforts have been cursory."[32] Numerous human rights reports documented continual abuses by CAFGUs throughout the 1990s.[33] One US State Department report, for example, stated:

> Civilian militia units or Citizens Armed Forces Geographical Units (CAFGUs) also committed extrajudicial killings. Organized by the police and the [Armed Forces of the Philippines] to secure areas cleared of insurgents, these nonprofessional units have inadequate training, poor supervision, and a propensity for violent behavior.[34]

It reported that other "vigilante groups or employees of contract security firms working with the authorities were also responsible for extrajudicial killings." In 2000, the CAFGU comprised of around 30,000 active participants.[35]

With the onset of the "War on Terror," these para-institutional relations were revamped alongside skyrocketing US military assistance and training in order to quell a resurgent communist insurgency, located primarily in Mindanao, as well as numerous Islamic insurrectionary forces such as the Moro Islamic Liberation Front (MILF) and Abu Sayyaf, as part of a commitment to "stability" operations and counter-subversive objectives. Consistent with para-institutional arrangements elsewhere, this has involved not only a fortification of the Philippine Armed Forces and its paramilitary appendages towards counterinsurgency and counter-terror operations, but has also relied on contracts with PMC to support it.[36] DynCorp, for example, was awarded a total of $164 million up until 2008 in multiple contracts to help train and support the Joint Special Operations Task Force in the Republic of the Philippines.[37] The Philippines also created a number of its own PMCs for foreign investment protection, such as guarding oil infrastructure and other areas of economic importance.[38]

Counterinsurgency and counter-narcotics efforts in Mexico also had significant parallel military structures, throughout much of the 1990s. Although the US and Mexico have long shared military-to-military ties, in which numerous Mexican military personnel were trained for internal policing, it was the rise of the Zapatista (*Ejército Zapatista de Liberación Nacional*—EZLN) insurgency in 1994–95, and an overlapping and increasing drug-trafficking problem, which prompted a closer military relationship between the two countries based on counterinsurgency and counter-narcotics.[39] The Mexican Armed Forces created a military development plan predicated on significant US military assistance, training, and equipment. This was primarily in response to the Zapatista insurgency in the Chiapas region that arose in opposition to neoliberal reforms and Mexico's integration into the North American Free Trade Agreement (NAFTA).[40] As part of these counterinsurgency efforts, around nine non-state paramilitary organizations emerged throughout Chiapas to counter these resistance efforts.[41]

Although Mexican paramilitary forces had local origins with connections to the ruling political party, the PRI (*Partido Revolucionario Institucional*), and elite land owners in the area, they also had significant links with the Mexican military, whose manuals and directives were "an almost literal translation of the U.S. Defense Department's *Field Manual Psychological Operations*" in describing a paramilitary option.[42] Much like other US-informed Central and South American counterinsurgency manuals of the Cold War, the Mexican *El Plan de Campaña* secretly issued to military commanders in Chiapas to counter the growing insurgency there, among other paramilitary recommendations instructed "secretly organizing certain sectors of the civilian population, small property owners and individuals with strong patriotism, who will be employed in the support of [these] operations."[43] The aims of paramilitarism were made pretty clear: "the command and coordination" of local public security troops and local ranchers "in the elimination" of the subversives and "the disintegration or control of social organizations." Although Mexican authorities and military leaders denied the existence of paramilitary organizations altogether, let alone their connection to them, declassified US diplomatic communications point to the direct support provided to non-state paramilitary groups operating against EZLN-sympathetic communities by Mexican authorities.[44] Media reports have also noted the explicit paramilitary-military connection in the promulgation of a counterinsurgency war against local movements (both armed and unarmed) opposed to the prevailing economic and political modes of development.[45]

Consistent with the Mexican military's counterinsurgency drive, paramilitary groups targeted

> ... EZLN activists and sympathizers, PRD [*Partido de la Revolución Democratica*] members, and officials or communities thought to be PRD or EZLN strongholds. Attacks have also been orchestrated

against human rights workers and advocacy groups, as well as civic and religious leaders suspected of being sympathetic to the "leftist" cause.[46]

Entire communities with declared links to the EZLN were targeted, as well as unarmed movements across the region. In one instance, in 1997, paramilitary groups initiated a prolonged terror campaign in towns throughout Chiapas, culminating in the Acteal Massacre on December 22 in which 45 people were killed by paramilitary gunmen.[47] According to Moksnes, the area experienced an increase of paramilitaries during the 1990s.[48] Similar paramilitary forces have been used in other areas of Mexico, such as in the state of Oaxaca and Guerrero, in order to contain incipient insurgent movements and deter other forms of "subversion."[49]

The Privatization "Revolution": The Consolidation of PMCs in US Statecraft

Within the context of the proliferation of neoliberal models and the expansion of global capital, PMCs became further integrated into US imperialist practices, serving to support and amplify US and transnational "stability" operations around the world. The private military and security industry grew exponentially in the wake of an increasing acceptance of a "privatization" in these areas, and PMCs became corporatized entities.[50] This provided an impetus for two inextricably intertwined developments. First, a wider variety of US government agencies contracted PMCs to conduct a wider variety of services. PMCs were effectively further incorporated into US and transnational stabilization efforts. Secondly, as MNCs extended their operations across the globe, many sought new ways to manage risks to their profits by liaising with US and local military forces as well as through extended networks of private coercive power.[51] By extension, PMCs and the private security sector became central to the lubrication of the global flows of capital and resources onto international markets, the protection of foreign direct investment, and conducive to the wider liberalized global order.

During the late 1980s and 1990s, US planners sanctioned outsourcing to PMCs to a greater extent and across a broader range of activities than ever before. Whereas during the Cold War PMCs primarily had performed mercenary tasks and logistical support roles for CIA covert operations, during the late 1980s and 1990s PMCs began to routinely conduct US military training exercises and a huge variety of support roles. US-based PMCs during the 1990s offered their military expertise to foreign countries around the globe; PMCs such as MPRI, SAIC, BDM International, Booz-Allen, and Vinnell, which had been operational in small-scale training, were now often hired to train entire militaries.[52] Vinnell continued to have an integral role in training the Saudi National Guard in counterinsurgency techniques and intelligence gathering to protect vital oil infrastructure and the Saudi Royal family from internal opposition.[53] Companies such as Booz-Allen and Hamilton trained the Saudi Marine Corps and consulted the Saudi Armed Forces Staff College, teaching "senior-level military skills," while O'Gara's and SAIC were involved with training Saudi private guards and the Saudi Navy.[54] The US and Angola contracted MPRI to train Angolan forces to quell a growing insurgency and protect lucrative mining businesses.[55] The contract, according to Silverstein, was meant to include "full-scale" training of the army and police and the notorious Rapid Intervention Police, which had a record of human rights abuses in attempts to deter the UNITA (*União Nacional para a Independência Total de Angola*) rebels.[56] Companies such as the Betac Corporation, which was previously involved with covert operations in the 1980s, were enlisted by United States Special Operations Command (SOCOM) to "assist US clients with internal security."[57] PMCs were also incorporated to a larger degree than ever before in a logistical fashion in support of US direct deployments of troops, such as in Iraq (in the first Gulf War), Somalia, Rwanda, Haiti, and South West Asia.[58] US military manuals began to include operational requirements for PMCs.[59] The legitimatization of PMCs during the 1990s entailed the incorporation of private military and security actors across all areas of US foreign policy, from intelligence and security, to training and operating technical equipment. The Department of Defense, the State Department, and USAID, among other US agencies, privatized non-core services. Washington facilitated these developments through a series of reforms designed to incentivize military and security outsourcing.[60]

However, contrary to a truly global market for force, throughout the 1990s and beyond the US formed a center (with the UK in close second) of the PMC market. Most PMCs around the world were US-based and most large PMCs hailed from the US (whereas UK and South African companies tended to be smaller in size).[61] Moreover, the US spent more on private companies than any other country during the 1990s and beyond.[62] By 1999, US annual expenditures on PMCs (contracts of all types) approached $50 billion, more than the entire defense budgets of all the other NATO members combined.[63] Some observers labeled the market for PMCs an American monopsony.[64] The centrality of US-based PMCs as primary drivers of the PMC market is indicative of the extent to which PMCs and their growth are products of processes within US foreign policy. Additionally, regulations were introduced to ensure that US-based PMC activities, whether contracted by the US or a foreign country, complied with US interests.[65] Companies such as SAIC, run by former Special Forces members, offered "military training and related assistance to foreign governments at the bidding of the United States."[66] Similarly, a former high-ranking DIA (Defense Intelligence Agency) officer observed that "The [privately run] programs are designed to further our foreign policy objectives ... If the government doesn't sanction it, the companies don't do it."[67] The US government exerted control over contracts bought between a third country and a US-based PMC via licensing agreements that grant permission to operate abroad. Additionally, US-based PMCs were usually owned and operated by ex-military personnel, and often hired ex-soldiers.[68] This underscored the transmission of US military specialists between the "public" and "private" realms in which former public employees go on to work in a private capacity for public goals.[69] These points help elucidate how the US government exercised significant coordinating control over US-based PMC contractors. As scholar Michael Likosky put it, "Even if we retain the term 'privatization', we should see privatization itself as created by public-private partnerships, rather than a move of activities from the 'public' and into the 'private' domain."[70] These points also help to highlight the ways PMCs were wielded as an instrument of US power rather than merely a result of an expanding "market for force." While an emerging global market for security and military expertise for stability was part and parcel of the growth of PMCs, another important component in the growth of a PMC industry (particularly in the US) was embedded in the ways that these actors had become central to the US's managerial role in the international system, its military-to-military ties for foreign internal defense in countries around the world, and for US-supported statecraft in the global South.

The proliferation of PMCs and continued promotion of localized paramilitary forces ushered in para-institutional complexes of force within the context of US-supported counterinsurgency stabilization efforts. These para-institutional complexes, built upon various nodes of outsourced coercive configurations, served to both "stabilize" certain social, political, and economic relations within states in the South and to provide security to transnational capital interests operational in areas beset by counter-hegemonic forces. PMCs were increasingly commissioned by the US (either hired by the recipient country or agencies of the US itself) to help host-nation security forces protect oil installations and other areas of economic interest, as part of a broader stabilization agenda.[71] Similarly, MNCs, often extractive industries, themselves often formed central conduits through which these para-institutional relations were formed. In this way, such para-institutional complexes including global and local "private" coercive actors during the 1990s increasingly took on roles as "investment enablers" within counterinsurgent contexts.

For example, coercive paramilitary forces in Peru during the 1990s were central to the stabilization of the liberalization of the Peruvian economy. The Peruvian military, with long-standing historical military ties with US counterinsurgency planners,[72] supported various pro-government civilian militias towards the military defeat of the *Sendero Luminoso* (Shining Path), the country's longest-lasting insurgency, from the 1980s, well through to the late 1990s.[73] In 1991, after his election, Peruvian President Alberto Fujimori legalized the use of so-called *Comités de Auto-Defensa* (CADs) and provided the means to arm peasant communities that were part of the civilian "self-defense" counterinsurgency program. Much like paramilitary initiatives elsewhere, despite the fact that they had diverse localized origins, "Counter-insurgency is the fundamental rationale behind [their] creation."[74] As Cynthia McClintock argued, amidst strengthening economic ties between the US and Peru and the increasing liberalization of the Peruvian economy, the support for paramilitary forces as part of a concerted counterinsurgency effort to defeat the *Senderos* created tensions for the two countries' relations due to concerns for human rights and democratic institutional development.[75] Despite this tension, US military assistance has continued (although not uninterrupted) ostensibly for counter-narcotics purposes throughout the post-Cold War period up until the present.[76] PMCs have also aided in stabilization efforts through direct participation in aerial surveillance, as well as in the protection of mines, oil fields, and other industries of international economic importance.[77] They have also been hired for counter-narcotics purposes in the Andean region, with the Peruvian military working directly with companies such as DynCorp under the auspices of US agencies.[78]

The expanding role of "private" actors in the stabilization of the Nigerian oil basin, and the increasingly central role that transnational corporations and PMCs played in assembling them, also serve as a good example of these para-institutional complexes in service of transnational capitalist interests. In the face of rising local unrest and insurgent targeting of oil installations, the US increased its military ties with Nigerian military forces, who in turn, in conjunction with transnational oil corporations in Nigeria, deployed global and local private actors such as PMCs (global) and militia forces (local) to protect their business assets and operations. These para-institutional arrangements for the internal stabilization of the Nigerian Delta have grown steadily since the 1990s.[79] In 2009, Shell reportedly provided $65 million to the Nigerian military as well as another $75 million on "private" security arrangements, including PMCs and militia or paramilitary forces.[80] In a similar case, alongside localized counterinsurgent militia forces operating on the ground who represented the sharp edge of coercive capabilities, Airscan was hired in Angola in 1997 to provide aerial surveillance against rebel attacks on oil pipelines.[81] In other words, there existed myriad relations between state forces and a collection of global and local para-institutional or "private" coercive actors operating in concert against centrifugal pressures that threatened the "stability" of functioning extractive businesses.

US agencies also covertly supported non-state military forces for a variety of smaller "irregular" operations and in some cases to mount an insurgency against unwanted regimes. For example, US Special Forces and the CIA were active in supporting and training insurgent and resistance forces, such as the Kuwaiti resistance and Kurdish rebel forces in Iraq as part of destabilization programs against Saddam Hussein. After the Iraqi invasion of Kuwait, as part of Operation Desert Shield/ Storm, the CIA and US Special Forces supported a Kuwaiti resistance guerrilla army to expel and hold back Iraqi forces.[82] These Kuwaiti resistance fighters mounted "sabotage" operations and other military efforts, and provided US and Coalition forces with intelligence on the Iraqi Army before direct US military involvement. Then within Iraq, the CIA supported a group of Iraqi exiles in the early 1990s "to plant bombs and sabotage government facilities" as part of a regime change agenda.[83] Alongside this, the US covertly supported other oppositional armed groups, such as Kurdish militias against Saddam's rule, with Congress approving a $40 million budget for these efforts in 1993.[84] Many of these forces were supplied via US PMCs.[85]

Colombia, Counterinsurgency, and Para-Institutional Complexes

Consistent with US imperialist strategies, US intervention into Colombia has long aimed at ensuring stability for overlapping US and transnational economic interests and maintaining a liberalized political economy in the region. US planners have sought to preserve Colombian policies beneficial to securing steady environments for investment, unfettered access to local markets by multinationals, and the fluid flow of capital. Left-wing insurgencies, principally the FARC (*Fuerzas Armadas Revolucionarias de Colombia*) and the ELN (*Ejército de Liberación Nacional*), and revolutionary unarmed social movements remained the principal threat to the desired order. This threat had to be addressed. US planners have repeatedly articulated the necessity of providing military assistance and training to Colombia in order to armor economic and political policies in Colombia that were deemed congenial to MNCs' access to Colombian resources and markets.[86] For instance, in 1999, the then Secretary of Energy Bill Richardson argued, "the United States and its allies will invest millions of dollars in two areas of the Colombian economy, in the areas of mining and energy, and to secure these investments we are tripling military aid to Colombia."[87]

US counterinsurgency assistance to Colombia increased in the 1990s and continued to rise with the implementation of Plan Colombia in 2000 and in light of the "War on Terror," the majority of which has been allocated to the Colombian military. Much of this aid, although ostensibly allocated for purposes of combating the illicit drug trade and then later terrorism, has consistently had a counterinsurgent orientation.[88] In turn, US-supported counterinsurgency in Colombia increasingly took on outsourced forms. Para-institutional forces became fundamental in preserving political economic configurations in Colombia favorable to global capital flows and stable investment climates. These forces targeted social and insurgent forces that threatened the desired liberalized order and provided security to MNCs' operations.

The Counterinsurgent Paramilitarization of Colombia

The nexus between US-supported counterinsurgency strategies and various right-wing militias, referred to as "paramilitaries" in Colombia, has been extensively documented.[89] Like elsewhere, paramilitarism in Colombia was not exclusively a product of counterinsurgency design, but rather had domestic and international systemic roots, molded by an amalgamation of processes within the neoliberal development of Colombia. The paramilitary phenomenon, according to numerous observers, developed out of the intersecting interests of large landowners, the political elite, including high-ranking figures in the government (even former President Uribe),[90] narco-traffickers, and multinational corporations; paramilitaries played a central role in the Colombian political landscape as part of a system of foreign capital penetration and expansion.[91] In her investigation of these dynamics, Hristov contended there are clear "links between US foreign policy, US capital interests, and the militarization of Colombian countryside," but equally important were the changing nature of "local power structures arising out of particular patterns of social relations."[92] In other words, sets of interests of local political and economic elites, transnational corporations, and large landowners were threatened by continued revolutionary action by insurgent forces jockeying for substantial political and rural reform. Out of efforts to preserve these interests various paramilitary groups were formed.

The Colombian state established the initial legal frameworks for paramilitarism in consultation with US generals to support counterinsurgent programs in the 1960s, as documented in previous chapters. Paramilitary formations in Colombia continued to expand and transform throughout the country's history. In the 1980s, groups such as *Muerte A Los Secuestradores* (Death to Kidnappers) were associated with the interests of drug cartels and large landowners. Additionally, according to several declassified documents, the CIA supported *Los Pepes* against the Medellin cartel towards the downfall of Pablo Escobar.[93] The leaders of these forces founded *Auto-Defensas Unidas de Colombia* (AUC), the United Self-Defense Forces of Colombia, a right-wing paramilitary umbrella organization that rose to prominence in the 1990s and early 2000s. The AUC were instrumental as a "para-extension to the Colombian state's coercive apparatus" in the conduct of a dirty war against the two main insurgent groups, the FARC and the ELN, as well as elements of civil society deemed subversive to the prevailing liberalized order.[94] The Colombian government also created programs such as Convivir in 1994, networks of informants and militias, as well as other auxiliary force units as part of plans to incorporate the participation of the civilian population in the state's counterinsurgency.[95]

As many have already argued, paramilitary forces, particularly during the 1990s, played a prominent role in attacking labor unions, workers' organizations, social movements, and insurgent forces that were deemed inimical or obstructive to the construction and maintenance of a functional capitalist state design.[96] MNCs operating in Colombia have also directly supported paramilitary groups to defend their immediate economic interests by paying them for security against insurgent damages. For example, between 1990 and 2004, Chiquita passed on millions of dollars to paramilitary forces and Convivir militia groups in payment for security and contribution to the overall counterinsurgent effort.[97] Companies such as Coca-Cola, Drummond, Chiquita, and others have been accused of relying on paramilitary forces to intimidate labor unions to ensure compliant workforces.[98] Paramilitary forces have also been the primary perpetrators of forced displacement in areas of economic importance, strengthening an already existing bond between violence and capital accumulation.[99] Carlos Castaño, the former leader of the AUC, himself proudly stated that his paramilitary organizations "have always proclaimed that we are the defenders of business freedom and of the national and international industrial sectors."[100] In simple terms, paramilitaries have constituted the tip of the spear in armoring processes of globalization against various forms of opposition from "below."

The relationships between the US and modern paramilitary forces such as the AUC are indirect, and there are limited (known) US direct links to these groups.[101] To the contrary, successive US administrations have publicly sought to limit or halt US military assistance to Colombia due to ongoing Colombian military-paramilitary links. Policymakers in Washington also supported the AUC demobilization process in the mid-2000s.[102] However, it is in the larger context of competing social and political movements and armed forces for the developmental future of Colombian politics and economics in which actors such as the AUC formed part of processes that served the interests of transnational capital. The AUC and other paramilitary forces were not surrogates directed by the US. However, such groups have functioned more indirectly as part of a counterinsurgency program in Colombia against "radical" elements of civil society jockeying for political and social change as well as insurgents.[103] They have served as a para-institutional mechanism through which US-supported counterinsurgency efforts have been prosecuted by the Colombian military.

PMCs in Colombia: The Privatized Means of US Intervention

Since the early 1990s, PMCs were also increasingly central to US counterinsurgency assistance to Colombia. PMCs facilitated US military assistance programs to Colombia, serving as a platform of operational support on which Colombian official military forces conducted the counterinsurgency war (with connections to paramilitary forces) against the FARC, the ELN, and, often, repression against social movements.[104] Additionally, multinational corporations have directly contracted US PMCs, contributing further to the overall "stabilization" agenda.[105]

The full extent of PMC activity in Colombia is not known, as the terms and conditions of these contracts as well as their operations were deemed private (as they are elsewhere).[106] Who contracts PMCs—whether the US governmental agencies, the Colombian government, or other international or private actors—further obscures reliable figures. It is clear, though, that a public-private infrastructure was gradually installed throughout the 1990s and early 2000s to support US military assistance to its Colombian partners. According to Colombian government records, the private security sector in Colombia, including those signed within US military assistance packages, grew by 360 percent between 1994 and 2007.[107] From 2002 to 2006, the value of US PMC contracts doubled, and roughly half of US military assistance since 2006 was implemented through PMCs (in 2006, $309.6 million out of $632 million).[108] The State Department and DoD contracted around 25 US-based PMCs for a variety of support roles to the Colombian military throughout the 1990s.[109] In addition, the Colombian government independently bought multiple multi-million dollar contracts, with full US State Department approval.[110]

The growth in PMC activity in Colombia was closely associated to increased levels of US military assistance and connected to a series of trends. Firstly, with multiple US military commitments across Latin America, Southern Command (SOUTHCOM) did not have sufficient capabilities to sustain significant US military presence in Colombia, and "employing PMCs enabled Washington to implement its Andean policy without undermining America's own military readiness."[111] In this context, PMCs served as "force multipliers." Secondly, these partnerships were a convenient way to obscure US involvement from Congress and the public.[112] US policymakers used private forces to bypass the varied policy restrictions Congress imposed on military assistance to

Colombia, including human rights conditions and troop caps.[113] PMCs also provided an avenue to circumnavigate potential political costs of US intervention. For instance, the State Department recorded 14 US citizens employed in Colombia as contractors had been killed between 1997 and 2003, five of them in 2003. General Nestor Ramirez, former Colombian Army commander and an ex-attaché to Washington commented, "Imagine if 20 American troops got killed here. Plan Colombia would be over."[114] Similarly, former US ambassador to Colombia, Myles Frechette highlighted: "It's very handy to have an outfit not part of the US armed forces, obviously. If somebody gets killed or whatever, you can say it's not a member of the armed forces."[115] In this case, rather than deny US complicity in these affairs, PMCs served to create a distance between the US and actions taken on its behalf.[116] Thirdly and finally, PMCs provided the flexibility, cost-efficiency, and expertise US policymakers demanded. Consistent with the advantages of privatization put forward by its advocates, US policymakers have claimed that outsourcing has been an expedient form allowing market forces to fulfill needed roles in an efficient manner.[117]

PMCs performed a range of activities from logistical and mechanical support to military training in Colombia.[118] For example, Lockheed Martin provided "logistics advisory, management, and professional services," and PAE Government Service conducted logistical support activities for the US and Colombian militaries.[119] In 2000, MPRI aimed to professionalize and develop the Colombian military's counter-narcotics/counterinsurgency capacity, as part of a previously devised "three-phase action plan" that involved "planning, operations (including psychological operations), military training, logistics, intelligence, and personnel management."[120] DynCorp provided 80 airplane and helicopter pilots and mechanics across the Andean region and in parts of Central America, earning at least $270 million between 1991 and 2001.[121] Other contracts provided in-combat intelligence and surveillance services. DynCorp and its subcontractors (one of which was EAST—a company involved in the Iran-Contra scandal)[122] were commissioned to conduct "eradication missions, training, and drug interdiction, but also participates in air transport, reconnaissance, search and rescue, airborne medical evacuation, ferrying equipment and personnel from one country to another, as well as aircraft maintenance."[123] Another company called Mantech International provided "complete technical support," which included intelligence on guerrilla locations.[124]

Such contracts brought PMCs closer to the front lines, pointing to levels of US involvement in the Colombian conflict that far surpassed what Congress had intended. In the mid-1990s, DynCorp flew Black Hawks and Huey II helicopters, trained Colombian pilots to use them, and maintained these aircraft under a $79 million contract granted to the Colombian military though US assistance programs. However, DynCorp pilots have reportedly surpassed their contractual limitations and have engaged in direct combat with the FARC, earning a local reputation for a willingness to "get wet."[125] Similar to Air America and other air-military contractors during Cold War-era paramilitary operations, DynCorp personnel were involved in direct engagements with Colombian guerrillas in efforts to extricate downed pilots in "search and rescue" operations.[126] Consequently, DynCorp became enmeshed in the conflict as the FARC and other guerrilla groups targeted their bases and refueling stations.[127] Up until 2003, Northrup Grumman piloted reconnaissance aircraft equipped with infra-red cameras to monitor guerrilla movements and drug-related activities.[128] A series of plane crashes and clashes with guerrilla forces resulted in the death of at least five contractors and the capture of three others by the FARC. The Colombian military rescued the hostages in the controversial *Operación Jaque* in 2008.[129] US planners and the Colombian government contracted PMCs to undertake a variety of tasks.[130]

In order to protect businesses from insurgent attacks, MNCs in Colombia increasingly hired PMCs throughout the 1990s.[131] Colombian government sources counted 573 companies that have their own security department mainly in the petroleum sector and other natural resource exploitation industries.[132] While this includes private security guards and basic installation protection, many have been military in nature, collaborating with US and Colombian military forces. For instance, a number of PMCs have provided operational and intelligence support to the Colombian military in protecting key oil facilities. US and Colombian court records demonstrate that in early 1998, Airscan, a US company hired by Occidental Petroleum to monitor insurgent activity to protect the Caño-Limon oil pipeline, was intimately involved in furnishing information that led to an air strike against a village called Santo Domingo.[133] Airscan shared information with the Colombian Air Force during meetings held at Occidental's facilities, in order to coordinate attacks against suspected insurgents who were allegedly in the area. The attack resulted in the deaths of 17 civilians. Similarly, British Petroleum and a slew of other affiliated oil companies operated alongside Colombian forces and were complicit in the creation of an informant network of local former Colombian soldiers to actively seek out insurgents in communities around the length of the Ocensa oil pipeline.[134] These allegations amount to the connections between the PMC activities gathering intelligence on local "subversives" alongside support to a parallel paramilitary formation to neutralize such civilian threat.

In this way, there is a symbiotic relationship between PMCs, the US-supported Colombian military-paramilitary nexus, and the transnational economic interests that they benefit. Links between PMCs, the Colombian military, and their paramilitary allies have been found in relation to Occidental Petroleum's monetary assistance to the Colombian Army's 18th Brigade, well known to have connections to paramilitary groups. Court proceedings against Occidental Petroleum alleged that money was funneled to the Colombian Armed Forces to protect the Caño-Limon oil pipeline whereupon Colombian forces, according to one of the claimants, "directly or indirectly (by supporting right-wing paramilitary groups), participated in numerous massacres of civilians and the disappearances, extra-judicial killings, arbitrary detentions, and beatings of social protestors."[135] Paramilitary forces are often given the unofficial go-ahead to attack suspected insurgents in efforts to exert further control and protection of oil installations and pipelines, acting on information garnered by PMCs provided to Colombian military sources. There have been further accusations of PMC-paramilitary links mediated in the context of US and Colombian counterinsurgency programs. According to a human rights organizer in Colombia, paramilitary forces "clear the ground" to prevent air-contractors spraying herbicides from being shot at by farmers or insurgents.[136]

There have also been cases of more direct PMC-paramilitary connections. For instance, in 1987, paramilitary groups and large landholders received training in counterinsurgency and "anti-subversive techniques" including lessons on how to "clean out" suspected members of guerrilla organizations from areas involved in oil and banana production from an Israeli military company called *Hod He'hanitin* (Spearhead Ltd.). The training allegedly took place on land owned by Texas Petroleum Co.[137] The Spearhead company's leader Yair Klein was later detained, but according to media reports, he claimed that such training to paramilitary forces was provided with the consent of the Colombian authorities.[138]

Similar charges emerged that British mercenaries have helped to train paramilitary groups.[139]

In summary, the prosecution of a US-supported counterinsurgent war in Colombia during the 1990s was increasingly delegated to para-institutional forces. This points to the need to avoid overly state-centric conceptualizations of US intervention into the Colombian conflict. It incorporated a broad network of para-institutional mechanisms. Many support and training activities of US assistance were contracted "outwards" in public-private partnerships and much of the "dirty work" of counterinsurgent violence was delegated "downwards" to paramilitary forces with disparate connections to the military, Colombian state, elite landowner interests, and the security requirements of transnational corporations. This elucidates the structural relations that form a nexus between the state and para-institutional actors and how they were generated and molded by the particularities of the Colombian context, the dynamics of the conflict and local elites. There is an organic fashion, dependent on the contours of the Colombian setting, by which this nexus is manifested within a US-supported counterinsurgency framework geared towards the stabilization of a liberalized Colombia.

Conclusion

There was significant advancement in the para-institutional means of extending US power in the post-Cold War period. These developments were most visible in the PMC sector. Primarily, US policymakers made military privatization official. PMCs became formal business entities that were fully and officially integrated into US military apparatus, taking on an increased variety of activities and to a greater degree. PMCs and private security forces became integral to US practices of statecraft, supporting and sometimes replacing US personnel in key US-backed stability operations. More broadly, PMCs evolved to form a global industry, with American companies and markets driving much of its expansion. Towards increased security measures, MNCs were often increasingly involved in US-backed stability programs, cooperating and coordinating with a complex of forces towards mutually beneficial goals. Public-private partnerships formed a centerpiece in US-supported counterinsurgency campaigns and a conduit through which processes of stabilization of capitalist socio-economic relations occurred.

Overall, there has been a growing military-paramilitary nexus in which a collection of para-institutional forces functioned toward US-backed counterinsurgency programs. The US increasingly outsourced many of its functions in the provision of counterinsurgency assistance "outwards" to PMCs, with such organizations often conducting the logistics, base and equipment maintenance, training exercises, and operation of technical hardware and flights. Simultaneously, governments in receipt of this assistance in allied states in the global South, with US training and/or acquiescence, colluded with militias, paramilitaries, and armed civilian groups, delegating central tasks "downwards." Meanwhile, these relationships often formed in the context of privatization of security and accelerated processes of capital accumulation in which local elites, large landowners, and transnational corporations sought ways to protect their operations, assets, and interests. Military-paramilitary relations also allowed the US to increase its global managerial commitments in policing the periphery, with military training and assistance to a higher number of allied states around the world and multiple interventions.

The example of US-supported counterinsurgency in Colombia underscored continuities in US foreign policy in the post-Cold War context, the continuation of counterinsurgency as an instrument of statecraft, as well as the acceleration of the multifaceted interests that informed these processes. During the 1990s, PMCs took on key tasks including training Colombian personnel, operating helicopters and flying intelligence reconnaissance airplanes among others. In addition, composite relations between state and paramilitary forces in Colombia formed part of a broader stabilization agenda. Colombian officials created paramilitary structures such as the Convivir program. However, the interests of wealthy landowners, drug traffickers, and elements of the elite ruling classes aligned to form military-paramilitary alliances with forces such as the AUC. MNCs and large agri-businesses contributed to these complexes via pursuing their interests through similar military-paramilitary relations and private security. Complexes of military and para-institutional military formations reconfigured the counterinsurgent articulation of power. As the next chapter covers, in the "War on Terror," these sets of para-institutional relationships have become truly global.

6

The War on Terror, Irregular Warfare, and the Global Projection of Force

With the onset of the "War on Terror," para-institutional means of extending US coercive reach expanded significantly. The military dimensions of Washington's role in the maintenance of a liberalized global order have been increasingly outsourced. Para-institutional complexes connected to US-allied power centers have become standardized, and flexible sets of relations that armor the fluid functioning of increasingly transnational capital flows have served as a coercive bulwark to maintain corresponding politico-economic arrangements in "unstable" states in the global South. The nexus between US official military forces and various para-institutional forces has become global in scope, with para-institutional complexes operational across the Middle East from counterinsurgent hotspots in Iraq and Afghanistan to conflicts in Yemen, Libya, Nigeria, Somalia, and Syria. They have been instrumental in the "stabilization" of investment climates and protecting vital infrastructure in diverse locations such as oil facilities in Nigeria, mines in Ghana, and critical supply routes in Iraq and Afghanistan, among many others.

Outsourcing and contracting to PMCs and alliances with local militias, warlords, and other paramilitary organizations, have become the *modus operandi* of US imperial statecraft. US military planners in conventional military branches (not just primarily the CIA and Special Forces) have embraced contracting to PMCs and working "with or through" local militias, paramilitaries and other actors. PMCs have conducted a huge range of services including logistical support, and base and aircraft maintenance, and have substituted for US personnel in military training missions, surveillance and intelligence gathering, transportation in the global rendition network, and front-line combat, among other roles. Similarly, as ostensibly new paradigms of "irregular warfare" dominated the blueprint for strategies of statecraft, officials in the Army, Air Force, and other agencies have incorporated support for local para-institutional forces into their mandates and operating procedures.

The delegation of force in US and US-supported counterinsurgency and unconventional warfare is consistent with long-standing Cold War coercive doctrines and practices. Lessons gained from US interventions and training programs in Latin America during the Cold War have now been applied across the globe, but in particular within unstable areas in the Middle East. The "world historical paramilitary movement" which the US helped to foment in Latin America and elsewhere during the Cold War era has now been replicated in the Middle East and across the world.[1] These developments were connected to the wider search for risk-averse ways of war.[2] Transferring military tasks (and the risks that go with them) to para-institutional forces has served to reduce the military demands not only on the US, but also on allied partner states in the global South.

The evolution of this nexus is connected to developments in global capitalism. The demand for security and stable sets of political and economic arrangements abroad has augmented alongside the growing transnationalization of capital. In this way, MNCs or transnational corporations have played an even more central role in the expanding coercive para-institutional complexes both in service of the security of their business operations and assets, but also in coordinating their operations with local governments and US military advisors. While operating in unstable or potentially dangerous areas, MNCs have coordinated their security arrangements with US-led military-paramilitary complexes. This has provided further layers to a lattice work of sets of relationships in the politico-military dimensions of US statecraft.

The "War on Terror" as US Imperialism

The onset of the "War on Terror" did not represent a significant change in US imperial grand strategies. Many analysts and US policymakers alike focused on the unilateral and preemptive application of US military power encapsulated in the Bush Doctrine and were quick to revert to the label "new imperialism" to describe the contours of US foreign policy in the "War on Terror."[3] While certainly an interesting dynamic that defined much of the Bush (George W.) administration's approach, the confluence of wider structural demands that had driven US foreign policy in the post-WWII era remained intact. Global capitalism advanced in its particular neoliberal forms. The growing transnationalization of a global economy has enhanced demands on the US as the primary guarantor of the stability of advantageous sets of political and economic policies abroad. US policies have continued to preserve the steady flow of resources onto global markets, to open new markets for transnational companies to sell their products (surplus absorption), and maintain new markets for trade, foreign direct investment (FDI) and wealth generation. The "War on Terror" has provided the context for which this role has manifested itself in specific forms.

The "War on Terror" not only constituted a commitment to combat the security threat posed by transnational terrorist organizations, such as Al-Qaeda, and localized counter-hegemonic forces in countries throughout the world, but has also included a continuation and expansion of long-standing strategies to provide the stability abroad conducive to a global liberalized order.[4] Largely replacing communism and socialism, in the Middle East anyway, political Islam (sometimes referred to as "Islamism") gradually emerged as a dominant counter-hegemonic social and political movement. As Saull succinctly put it, across the Middle East and elsewhere, it "is no longer the color red that dominates banners on popular demonstrations but the green of Islam."[5] With this, military analysts in the Bush administration were quick to conceptualize the oper-ational reality of the "War on Terror" not only as a war against "terrorist" organizations, but as a series of localized counterinsurgency campaigns and unconventional warfare operations in various parts of the world threatened by "destabilizing" internal forces.[6] The US renewed its efforts to counter political developments abroad that it deemed inimical to US and transnational interests. This included, but was by no means limited to, countering political Islamist movements.

Internal US documents point to this line of thinking among senior policymakers in the Bush administration at the initial stages of the "War on Terror." For example, the then Secretary of Defense Donald Rumsfeld questioned the nature of the "War on Terror" and the use of these words to describe US strategy. He instead focused on how the current threat of Islamist aspirations is not just a material security threat, but rather a political and ideological challenge.[7] History had not "come to an end" as boldly claimed at the end of the Cold War. Yet, Rumsfeld and others in the Bush administration quickly understood that by declaring a war on "terrorism" more generally, they could declare a war against whomever they deemed inimical to the stability of the US-led liberalized global order.

US national security planning documents point to the continuation of the underlying structural drivers of US foreign policy. The 2008 *National Defense Strategy* stated directly that

> Since World War II, the United States has acted as the primary force to maintain international security and stability … Driving these efforts has been a set of enduring national interests and a vision of opportunity and prosperity for the future. US interests include protecting the nation and our allies from attack or coercion, promoting international security to reduce conflict and foster economic growth, and securing the global commons and with them access to world markets and resources.[8]

Therefore, it posited, as a fundamental part of defeating terrorist networks, the preservation of specific forms of "stability" in the pursuit of US and transnational economic interests has taken center stage in US defense and security strategies. Many other national security and military documents contain similar statements regarding these priorities.[9] In effect, the US has continued to armor processes of globalization around the world working in conjunction with local elite ruling strata and state forces. While military-to-military ties with allied states is a primary channel through which this is operationalized, the US has often resorted to more interventionist modes of statecraft to help police peripheral states.

A prominent example of these dynamics is the invasion of Iraq. The logic driving US intervention and the specific counterinsurgent forms it has taken upon occupation not only pivoted on the removal of a leader perceived to be an immediate threat to US national security, but also centered on the transnationalization of the Iraqi political economy and its integration into the global capitalist system more broadly, with the protection of the continual flow of oil into international markets a top priority.[10] As many have argued, helping to steady international oil prices, diversification of global oil supplies, and guaranteeing the continued flow of petroleum and other resources onto international markets were a central consideration in the overthrow of Saddam Hussein's regime and the transformation of Iraqi political economy. As extensively documented elsewhere, MNCs and transnational corporations flooded into Iraq as part of the US-led approach to the nation-building project that ensued.[11] The US led efforts to contain or pacify disparate counter-hegemonic movements and processes. While predominantly manifest in measures against an anti-occupational insurgency, this included the pacification of other forms of "resistance." Trade unionists and workers' organizations appealing against the privatization of key industries, including oil, were met with force and intimidation coupled with policies enforcing minimal union activity.[12] As Herring and Rangwala have succinctly put it, US economic policies in Iraq as well its counterinsurgency initiatives against resistance, have worked to "open up the Iraqi economy," in accordance with a "US version of the neoliberal model."[13]

Irregular Warfare and Agents of US Imperialism

US military planners in Washington declared "irregular warfare" the cornerstone of US statecraft and fundamental to US strategies in the "War on Terror." Following the release of new counterinsurgency blueprints for Iraq and Afghanistan, the 2006 *Quadrennial Defense Review* report asserted that the "War on Terror" "requires the U.S. military to adopt unconventional and indirect approaches."[14] The DoD issued *Directive 3000.07* (initially in 2008, but then again updated and issued in 2014) outlining how "irregular warfare" and "working through irregular forces" is "as strategically important as traditional warfare" in order to "extend US reach."[15] "Irregular warfare" referred primarily to counterinsurgency and unconventional warfare but encompassed a wealth of coercive strategies designed to counter forces deemed to threaten the desired "stability" abroad.[16] It is a revised military umbrella term similar to "low-intensity conflicts" (a military umbrella term prevalent throughout the 1980s) and constitutes a continuation of Cold War forms of statecraft.[17]

The "irregular warfare" paradigm emphasized working "with or through" various "irregular forces."[18] US military documents defined these "irregular forces" as "individuals or groups of individuals who are not members of a regular armed force, police, or other internal security force," including "paramilitary forces, contractors, individuals, businesses, foreign political organizations, resistance or insurgent organizations, expatriates, transnational terrorism adversaries, disillusioned transnational terrorism members, black marketers, and other social or political 'undesirables.'"[19] Manuals used to train US military personnel and foreign official forces emphasized that this might pose risks as activities "frequently involve the irregular forces of non-state armed groups with questionable personalities and motives."[20] Nevertheless, according to these documents the expected utility of operating "with or through" "irregular forces" was "multiplying US power"[21] without direct participation or commitment of US forces, thus providing a "perception of USG [US government] restraint" and preserving an image of US non-intervention.[22] In effect, they allowed military planners, according to US irregular warfare doctrine, to "extend US reach into denied areas and uncertain environments."[23]

Under this irregular warfare umbrella, counterinsurgency and unconventional warfare training manuals and other documents identified militias, civilian armed groups, and other paramilitary forces as central to US strategies. Unconventional warfare training manuals throughout the 2000s and 2010s delineated US participation in mobilizing and supporting insurgencies of "irregular forces," including guerrillas, cooperative civilians, militias, and other armed forces to disrupt or overthrow "unfriendly" regimes.[24] In 2008, the DoD defined unconventional warfare as

> … a broad spectrum of military and paramilitary operations, normally of long duration, predominantly conducted through, with, or by indigenous or surrogate forces that are organized, trained, equipped, supported, and directed in varying degrees by an external source. It includes, but is not limited to, guerrilla warfare, subversion, sabotage, intelligence activities, and unconventional assisted recovery.[25]

Another report further outlined that "In unconventional warfare, US forces foster and/or support insurgencies against an established government. These operations are characterized by their low visibility, covert, and clandestine nature."[26]

Similarly, there is a general understanding throughout US counterinsurgency doctrines that militias, "civilian defense forces," "home guards," and other "paramilitary" organizations should be leveraged towards the counterinsurgency cause either directly by US forces or indirectly via training local state forces to do the same.[27] For example, a 2009 US military training handbook title *Tactics in Counterinsurgency* advised that paramilitary forces augment state capabilities: "if the HN [Host Nation] security forces are inadequate, units should consider hiring a paramilitary force to secure the village or neighborhood."[28] One 2004 US military document on "Foreign Internal Defense" described paramilitary forces as playing a fundamental role in counterinsurgency operations, intelligence gathering, and psychological operations.[29] Another outlined that in US counterinsurgencies, "local paramilitary forces—including home guards, village militia, and police auxiliaries—are mobilized or created, organized, and trained as reserves."[30]

The stated strategic rationales for these tactics remained intact from their Cold War ancestry. One US military document argued that Special Forces were designated specifically to "select, organize, and train paramilitary and irregular forces," to "serve as force multipliers in the pursuit of U.S. national security objectives with minimum U.S. visibility, risk, and cost."[31] One manual used to train foreign forces instructed that paramilitary formations should be created to augment official state personnel: "If adequate HN [host nation] security forces are not available, units should consider hiring and training local paramilitary forces to secure the cleared village or neighborhood. Not only do the members of the paramilitary have a stake in their area's security, they also receive a wage."[32] Much like Cold War-era conceptualizations of "counter-organization," US planners continued to view paramilitarism as an expedient political tool to garner active participation from local citizens in the counterinsurgency effort. For example, one Wikileaked 2004 Special Forces manual contained a section on mobilizing civilians into self-defense forces in order to garner support for the local government and deny civilian safe havens for insurgents. It contained identical tactical rationales for their mobilization as Cold War counterinsurgency instructional handbooks. It explained that the "average peasant is not normally willing to fight to his death for his national government … The CSDF [civilian self-defense force] concept directly involves the peasant in the war and makes it a fight for the family and village instead of a fight for some far away irrelevant government."[33] In a review of militia forces in Afghanistan, one US military report found that such civilian defense forces "knew the ground better and could more easily spot something that was out of place or suspicious," making US forces much more effective in their counterinsurgency duties.[34] Similar rationales are provided in other manuals.[35]

This drive to strengthen "irregular warfare" capabilities globally entailed simultaneously expanding US capacity to directly support "irregular" forces as well as increasing the provision of counterinsurgency assistance and training to "friendly" or allied states.[36] Towards this end, CIA and Special Force paramilitary mandates and budgets were significantly expanded in the early 2000s.[37] For example, at the onset of the

"War on Terror," Donald Rumsfeld augmented DoD "special operations" capabilities via the US Special Operations Command (SOCOM) from $2.3 billion for FY 2001, to a peak of $12.1 billion in FY 2011, $10.4 billion in 2013, and totals around $10 billion in 2015 as well as increases in personnel and number of missions.[38] Subsequently, Special Forces have been deployed to over 140 countries around the world conducting specific missions and liaising with local partners and para-institutional forces.[39] In addition to this, section 1208 of the Ronald W. Reagan National Defense Authorization Act for 2005 granted the DoD around $25 million annually in order to provide "support to foreign forces, irregular forces, groups, or individuals engaged in supporting or facilitating on-going military operations by United States special operations forces."[40] Additionally, US military assistance and training programs to foreign allied states have increased in the "War on Terror," and have reached record levels of expenditure and scope in terms of numbers of military personnel trained, towards internal policing and "stability" operations.[41] Many training programs were increasingly outsourced to US-based PMCs (sometimes with the recipient state doing the contracting and paying) as a way to extend US influence and also thereby falling outside the remit of official budgets and Congressional restrictions and oversight.

While the CIA and Special Forces have remained the principal channels through which the US has forged alliances with para-institutional groups, other departments and agencies have adopted "irregular warfare" paradigms and integrated these practices into their mandates. Rather than the preserve of the CIA, Special Forces, and other more covert agencies, concerted efforts have been made to institutionalize "irregular" modes of coercive statecraft across US departments. Similarly, military studies proliferated on topics such as "irregular warfare," "surrogate warfare," "indirect approaches," and other terms used to describe relying on para-institutional actors, including militias, paramilitary groups, and private military forces to complete US objectives abroad.[42] For example, US military colonels articulated that the US military should embrace its "culture of Irregular Warfare—advising, liaising, training, leading and operating closely with local tribal levies, militias and other non-state forces."[43] This drive to strengthen "irregular warfare" capabilities placed the projection of US military power through para-institutional forces firmly at the forefront of US tactics.[44]

The PMC Explosion and US Global Force Projection

To support US military presence around the world and to buttress US interventions and statecraft, US policymakers have accelerated the delegation of non-core military services to PMCs. Outsourcing to PMCs in the "War on Terror" reached unprecedented scale in terms of their sheer breadth and scope of their activities in service of US foreign policy objectives. The DoD has spent more than half its budget on PMCs over the course of the "War on Terror."[45] In 2015, the DoD spent $274 billion on contracts in total.[46] These figures do not include those outlays of other US agencies, such as the State Department or the CIA. Nor does it include contracts paid for by countries either with their own funds or recycled US assistance. The actual numbers of contractors working in the service of US-led stability programs are much higher. PMCs are now a common feature of US military deployments and assistance programs and are active on a global scale.[47] PMCs have taken on an increasing variety of roles in US global military power projection. This is across the spectrum in logistics, the construction and maintenance of US military bases, operating technical equipment, flying planes and drones, conducting interrogations—indeed, almost anything except making major strategic decisions.

PMCs have bolstered US imperialism in the "War on Terror" in two primary ways. First, US planners have increasingly outsourced to PMCs to enhance official US military capabilities. In the "War on Terror," outsourcing to PMCs has become essential to maintaining global US military presence and has become integral to the US military machine straddling the world. They allow the US military to maintain a presence around the world and deploy US forces to multiple areas simultaneously if needed.[48] Second, PMCs have also increasingly supplemented and often replaced official military personnel in a variety of front-line tasks, forming part of para-institutional complexes. PMCs are central to US coercive strategies of statecraft, working in tandem with US forces, and local governments and armed forces, as well as "irregular" forces contributing to US and US-supported counterinsurgency and unconventional warfare worldwide. PMCs now offer military training of foreign forces, operate high-tech weaponry and drones, collect and analyze intelligence, secure vital infrastructure and geo-political natural resources, and even perform mercenary-type front-line tasks.[49]

US counterinsurgency and unconventional warfare training to foreign forces with much of the global South (as well as other areas of military activity) have increasingly taken on outsourced forms.[50] PMCs have trained foreign forces in lieu of US trainers in countries such as Liberia, Malawi, Nigeria, Colombia, Saudi Arabia, Iraq, Afghanistan, and many others.[51] Drones and unmanned aerial vehicles (UAVs) have become central to US counterinsurgency and counter-terror in Afghanistan, Pakistan, Somalia, and Yemen among others, in the implementation of a kill-list of high-value targets. In turn, the US has increasingly outsourced the operation of drones for intelligence reconnaissance and missile strikes. For example, the first deployment of the Global Hawk drone system consisted of 56 contract operators out of a team of 82 for drone "combat-type operations."[52] Vinnell Corp. and Airscan have operated aerial surveillance in countries such as Iraq, Colombia, and Angola.[53] An investigation of US military documents revealed that "America's growing drone operations rely on hundreds of civilian contractors, including some—such as the SAIC employee—who work in the so-called kill chain before Hellfire missiles are launched," and who were involved in various incidents of civilian deaths from drone strikes. US policymakers have also hired PMCs to analyze information collected from drones and satellites over Afghanistan, with SAIC winning a $49 million contract for this purpose.[54] PMCs conduct interrogations and other intelligence-gathering missions/functions.[55]

PMCs have also played an important security function as part of US stabilization programs. For example, PMCs have been operational in protecting oil facilities[56] and security services (such as protection services for Paul Bremer, head of the Coalition Provisional Authority (CPA) in Iraq, in 2006).[57] This also forms part of an escalating trend in which private armies have been contracted, in connection to or in part of US and local counterinsurgency campaigns, to insulate lucrative resource extractive business from the damaging effects of conflict and deliberate attempts by rebels to disrupt their operation.[58] Erinys, for instance, has experience in protecting the Ahanti gold mine in Ghana and MPRI's agreement to professionalize and enhance Equatorial Guinea's military capability to protect key oil facilities is another good example of this trend.[59] Vinnell has continued to secure contracts training the Saudi National Guard (SANG) in maintaining the Saudi royal family's grip on power in the country as well as, in conjunction, protecting oil pipelines and infrastructure.[60] PMCs have also conducted combat roles in US counterinsurgency and unconventional warfare. The *New York Times* reported that the CIA hired Blackwater to locate and kill Al-Qaeda leaders as part of a clandestine assassination program.[61] In a throw-back to the Cold War era reign of Air America and Civil Air Transport, private air-contractors formed the backbone of US-led global system of extraordinary rendition. The entire transportation network that made up the extraordinary rendition program was conducted by private airliners with connections to the CIA and DoD on contract to fly detainees around the world to secret locations and military bases, often ending at the Guatánamo Bay prison.[62]

The Global Proliferation of US-Backed Para-Institutional Complexes

With these two separate but related developments, there has been a global proliferation of US-backed para-institutional forces. Complexes of para-institutional forces have become a centerpiece of US power projection and essential to US military presence and US-backed operations around the world. US strategies to stabilize liberalized politico-economic orders abroad have been implemented via paramilitary extended means in which significant state relationships with para-institutional forces have developed. This has involved composite relations with collections of para-institutional forces in order to project US power around the world and as part of localized politico-military statecraft. These relationships have helped shape para-institutional complexes, a nexus between state military forces and paramilitary organizations and private armies, towards order making.

The US-led intervention in Afghanistan is one example. The entire project of transforming the social, political, and economic order in Afghanistan (Operation Enduring Freedom) has been underpinned by complexes of force that connect state-led processes to para-institutional formations of coercion. Towards the overthrow of the Taliban regime in Afghanistan, for example, the CIA and Special Forces mobilized anti-Taliban guerrilla forces in September 2001, primarily the Northern Alliance and other forces to the north of the country (Operation Jawbreaker is an example) and various Pashtun tribes in the south.[63] The US depended on these allies-for-hire for intelligence gathering, conducting "sabotage" operations, and marking targets for US aerial bombardments. Many groups received cash handouts (sometimes receiving cash payments of up to $1 million) from the CIA and Special Forces for collaboration in destabilizing the Taliban regime in preparation for the eventual arrival of US forces, as well as leading counter-terror operations in search of Al-Qaeda operatives and Osama Bin Laden.[64] This included, according to some estimates, the delivery of "767 tons of supplies and $70 million, sufficient to equip and fund an estimated 50,000 militiamen."[65] According to reporter Bob Woodward, "In all, the U.S. commitment to overthrow the Taliban had been about 110 CIA officers and 316 Special Forces personnel, plus massive air power."[66] The operation, according to US military reports, was mounted much faster and more efficiently, and it was cheaper hiring local collaborative forces, such as the Northern Alliance, than might have been solely via the deployment of US forces: "By spring 2002, 6 months after the assault, the United States spent only $12 billion and lost about a dozen American lives."[67]

Immediately following the extirpation of the Taliban from the halls of the Afghan government, US military planners mounted a US-led Coalition counterinsurgency program to pacify the insurgency composed of the remnants of the Taliban, a collection of Haqqani tribal networks, anti-occupation dissidents, and elements of Al-Qaeda. The long-term objectives constituted a nation-building exercise and therefore included the creation and development of an Afghan government and its security forces (the Afghan National Army—ANA, and Afghan National Police—ANP). Yet, to do so involved developing networks of para-institutional forces towards the stabilization of a new order with a "pro-US" political and economic orientation. In US-led efforts to stabilize a post-Taliban order in Afghanistan, US officials applied what they referred to as an "indirect approach."[68]

Domestic Afghan political and power dynamics and the interplay between the consolidation of state power, local warlords, militias, and other tribal factions, which have all jockeyed for their own domains of control, significantly shaped the contours of US-militia collaboration. The "indirect approach" applied was emblematic of the intersection of US para-institutional stabilization strategies and local political power structures. In Afghanistan, warlords and militias have historically occupied influential roles in local governance and in processes of state formation.[69] In the context of the fractured political landscape and the power vacuum created after toppling the Taliban regime, rather than mobilize its own militia groups, US counterinsurgents worked with the warlords' political power structures.[70] In attempts to consolidate state institutions, the US and Coalition forces and the newly established Afghan government (headed by Hamid Karzai) conceded to and supported regional warlord power brokers, which in many cases impeded the development of centralized bureaucratic institutions and a strong security apparatus.

Counterinsurgency experts converged on the idea of harnessing the power of tribal factions, producing a torrent of articles and workshops extolling the virtues of incorporating the various tribes and warlords into the counterinsurgency strategy in Afghanistan.[71] Leaked US military command communications detailed a "Tribe-Centric" strategy which aimed to co-opt and cooperate with local warlords and militias.[72] US command often paid warlord leaders on an ad hoc basis to participate in military operations against insurgents.[73] In other cases, US agencies contracted out tasks to such groups. According to some estimates, the Coalition paid up to $10,000 a month in salaries to select warlord leaders for their participation in counterinsurgency. US agencies doled out cash payments, and even Viagra as compensation in some cases, for the loyalty of local collaborators.[74] Obama's 2009 Defense Appropriation Bill allocated $1.3 billion towards financing the "reintegration" of Afghan warlords into a semi-official militia force.[75] Similarly, under a so-called "Community Defense Initiative," US and Afghan agencies began to co-opt tribal forces, integrating them into the state-led counterinsurgency drive.[76] However, attempts to incorporate militias and warlords into state structures ("Disarmament, Demobilization, and Reintegration" (DDR) programs in 2003–05) failed in large part due to the demand placed on them for their counterinsurgent capacity.[77]

In addition to these military-warlord collaborations, US planners mobilized new counterinsurgent paramilitary forces.[78] One 2006 US counterinsurgency manual written specifically for Afghanistan cited the importance of constructing local "self-defense units," "paramilitary forces," and "friendly guerrilla" armies, in order to separate the insurgents from their civilian support base.[79] US strategists helped to establish the Afghan National Auxiliary Police (ANAP) program (between 2006 and 2008), mobilizing civilians and former militia members to supplement the official military.[80] US policymakers contrived other "self-defense" strategies.[81] US forces in conjunction with the Karzai government formed the Afghan Local Police (ALP), another similar civilian militia to "secure local communities and prevent rural areas from infiltration of insurgent groups."[82] The program relied on civilian recruits vetted by local police and trained by the US Special Operations Forces. Although the Afghan government was initially reluctant to mobilize the ALP, for fear of empowering local warlords and undermining state consolidation of power, by March 2012, the ALP counted 12,600 personnel with plans to expand the paramilitary force to over 30,000 members.[83] Despite an ostensible defensive capacity, they have reportedly committed wide-scale human rights abuses in search of suspected insurgents.[84] Multiple reports also highlighted links between US agencies and the Afghan security services to militias (such as the Shaheen unit) conducting night raids and extra-judicial killings of civilians suspected of supporting the Taliban.[85]

US stabilization strategies in Afghanistan also relied heavily on PMCs. PMCs have been instrumental in training and developing the new Afghanistan government's security and military apparatus by advising and mentoring the Ministry of Defense, the Afghan National Army, and national police service (the ANP).[86] PMCs also took on central responsibilities in the Afghan security apparatus. For instance, PMCs were hired as bodyguards for President Karzai, have been embedded in the Afghan military as advisors, conducted counter-drug operations, and guarded vital roads and other infrastructure.[87] Outsourcing and contracting to PMCs also provided the basis for para-institutional security complexes embedded within the broader context of a highly politicized process of US-directed statecraft and order-making. For example, a US House of Representatives investigation, titled *Warlord Inc.*, into the DoD's "Host Nation Trucking" multi-million dollar contracts, which outsourced the transportation US military supply lines and the protection of key roads, found companies working for US military logistics between 2007 and 2009 hired PMCs who in turn subcontracted security and other operations to local warlords and militia forces. This report detailed how US-hired security PMCs subcontracted much of their work by paying local militias and warlords, noting that the "principal private security subcontractors on the HNT [Host Nation Trucking] contract are warlords, strongmen, commanders, and militia leaders."[88] Another US government inquiry into the oversight of PMC activities in Afghanistan revealed chains of subcontracting over which the DoD had no control, leading to massive complexes of para-institutional formations working towards US stabilization programs.[89] In effect, US planners and contractors paid local PMCs run by warlord factions to provide security for various US installations and strategic infrastructure. This extended to security for Bagram and Shindand airbases, and various US military camps, as well as US military convoys and supply routes.

While simultaneously building the Afghan state and its military and police forces, the US presence in Afghanistan has been mediated primarily through PMCs and local warlords. According to official estimates, PMCs made up around 75 percent of US forces in Afghanistan.[90] By 2016, estimates were that there were three times as many PMCs as US personnel in Afghanistan.[91] None of these estimates include how US-based PMCs hired by third countries or MNCs contribute to US strategies.

As the Afghan conflict rumbled on into 2018, the Trump administration's recommendations for US involvement in the pacification of the protracted insurgency and the stabilization of Afghanistan have followed the well-established practices of limiting US direct intervention in favor of outsourcing and local collaboration. Erik Prince, the founder of Blackwater (which changed its name to Xe and then to Academi), proposed to limit the costs of US wars in Afghanistan by contracting out US operations to a 5,000-man privatized army.[92] While this specific proposal has been rejected, the Trump administration has been on course to further develop outsourced means of extending US influence in Afghanistan. Meanwhile, US military advisors introduced plans to reinvigorate US-backed self-defense militias in November 2017, reorganized through a new program called the "Afghan National Army Territorial Force."[93]

US-led intervention for the restructuring of Iraq's political economy followed similar patterns. Before and during the 2003 US invasion, in accordance with US unconventional warfare practices, the CIA and US Special Forces covertly orchestrated an insurgency to help topple Saddam Hussein's rule.[94] The CIA spent millions of dollars in training the Scorpions to "foment rebellion, conduct sabotage, and help CIA paramilitaries who entered Baghdad and other cities target buildings and individuals."[95] CIA and Special Force teams liaised with various Kurdish militia insurgent forces and mobilized and trained a collection of paramilitary groups to wage an insurgency against the Iraqi government in preparation for the arrival of US troops for the invasion.[96] PMCs supported the US military invasion force in a number of contracts, mainly towards base maintenance and operation of technical equipment.[97]

After removing Saddam from power, US planners devised and implemented a US-led paramilitary counterinsurgent plan towards the stabilization of a transnationalized economy in Iraq. US-led counterinsurgent strategies in Iraq depended heavily on local collaboration and the formation of militia forces to operate in conjunction with the US and the developing official Iraqi armed forces throughout the campaign. In many instances, the US directly mobilized militia assets. According to reporter Robert Dreyfuss, in November 2003, the US approved a covert budget of $3 billion towards the "creation of a paramilitary unit manned by militiamen associated with former Iraqi exile groups" to root out leftover Baathist sympathizers.[98] The funds for such a plan were to be channeled through the CIA to create paramilitary units and entice tribal Sunni leaders, exiled members of the Iraqi elite, and various militias to join the counterinsurgency effort. Then in 2005, media outlets reported on discussions of a so-called "Salvador Option," memorializing US assistance to El Salvador's death-squad networks during the 1980s. The application of a "Salvador Option" in Iraq, according to *Newsweek*, meant that the US would send "Special Forces teams to advise, support and possibly train Iraqi squads, most likely hand-picked Peshmerga fighters and Shi'ite militiamen, to target Sunni insurgents and their sympathizers."[99] While US military personnel with experience in counterinsurgent wars in Latin America were posted to Iraq, media and academic commentators pointed to the similarities between US tactics in Iraq and those employed in El Salvador and Vietnam.[100] During a Congressional hearing regarding Special Force operations, US Army Lt. Gen. William Boykin compared programs in Iraq to US paramilitary activities during Vietnam: "Killing or capturing these people is a legitimate mission for the department. I think we're doing what the Phoenix program was designed to do, without all of the secrecy."[101]

US planners, in conjunction with a newly formed Iraqi government, also formed semi-official elite paramilitary forces (in 2005–09). For example, the "Special Police Commandos" reported directly to the Iraqi Ministry of the Interior and was supervised by US military advisor James Steele (who had previously headed the Military Advisory Group in El Salvador during the early 1980s).[102] The "Wolf Brigade" was a similar semi-official paramilitary program with which US and Iraqi forces conducted counterinsurgency raids.[103] These forces were designed, according to one academic observer, "as an extrajudicial method of capturing and killing suspected Sunni insurgents," as well as those social and political forces deemed subversive to the US and Coalition political and economic program.[104] They were often accused of death-squad activities and human rights violations against Iraqi civilians, including disappearances, assassinations, torture, and unlawful detentions in pursuit of suspected insurgents and their sympathizers.[105] Other victims of Iraqi paramilitary violence consisted of various specific political and social forces in opposition to economic policies favoring privatization and neoliberal structural reforms. For instance, Max Fuller reported that trade unionists and workers' organizations appealing against the privatization of key industries were met with force and intimidation, coupled with state policies in favor of privatization and minimal union activity.[106] Similar counter-resistance was launched against members of the Iraqi Federation of Oil Unions opposing the privatization of the Iraqi oil infrastructure, with the Iraqi government issuing arrest warrants for union leaders.[107] According to a 2010 Amnesty International report, political activists, human rights workers, and journalists opposing occupation and critical of the Iraqi government were the subject of reprisals from paramilitaries.[108]

Comparing them to patterns of death-squad activity in El Salvador, former UN Human Rights Chief in Iraq John Pace confirmed findings that victims recovered from mass graves showed signs of torture, having been blindfolded, handcuffed, and executed with gunshots to the back of the head.[109] The media reported a flurry of dead bodies disposed of in rivers, abandoned buildings, and other public places to send a message to those supportive of the anti-occupational insurgency.[110] Human Rights Watch commented that "Every month, hundreds of people are abducted, tortured and killed by what many believe are death squads that include security forces. To terrorize the population, the killers often dump the mutilated corpses in public areas."[111] Some of their interrogations of suspects were broadcast on national TV show called "Terrorism in the Grip of Justice."[112] A US advisor to the Iraqi government told reporters: "The only way we can win is to go unconventional. We're going to have to play their game. Guerrilla versus guerrilla. Terrorism versus terrorism. We've got to scare the Iraqis into submission."[113]

Meanwhile, the Coalition forces consisted of a significant number of global PMCs at almost every level of US-led stabilization efforts. The provision of military and police training to Iraqi forces for internal policing was delegated to private companies.[114] In 2003, Vinnell, a subsidiary of Northrop Grumman Corporation, was awarded a $48 million contract to train the first nine battalions of the new Iraqi Army.[115] The majority of Iraqi police forces have also been trained by DynCorp with around $2.5 billion in contracts between 2004 and 2011.[116] US policymakers also contracted PMCs to train militias, paramilitary commando squads, and other para-institutional forces. US Investigations Services won a $64.5 million contract in 2004 to train local militias such as the Emergency Response Units and some elements of the Sons of Iraq (for more of which, see below).[117] In Iraq, multiple companies have been contracted for aerial surveillance and the provision of real-time intelligence to US and Iraqi forces.[118] Airscan, for example, won a $50 million contract from the US military in March–December 2011 to provide "real-time over-target full-motion video from commercial manned airborne surveillance platforms for Iraq-wide air surveillance support."[119] Companies such as Titan and CACI conducted interrogations and were implicated in the Abu Ghraib torture scandals.[120] The CIA also reportedly hired Blackwater and other PMCs for covert operations and central combat roles in the early phases of US intervention into Iraq.[121]

PMCs and militia groups have protected vital infrastructure and installations as part of broader counterinsurgency goals, serving as investment enablers.[122] For example, according to its website, Erinys International "created and deployed over a period of 18 months a guard force of over 16,000 Iraqi national security guards (both fixed site and mobile) protecting 282 key oil infrastructure sites, including strategically significant oil and gas pipelines."[123] Aerial surveillance was subcontracted to Airscan to monitor the pipelines at night. This constituted only a small part of a larger "Facilities Protection Services" program established by Bremer's CPA in 2003 to protect Iraqi government buildings, public facilities, and critical infrastructure, as part of an overall stabilization plan, some of which was also spearheaded by local militia forces.[124]

Due to the myriad of contractors entering the fray, the Pentagon established a mechanism to coordinate between all the PMCs operational in Iraq and the US military called "Project Matrix." This itself was contracted out to Aegis Defense Services, a British PMC founded by, among others, Tim Spicer, former CEO of PMC Sandline International.[125] The central aim of the $293 million contract signed in 2004 was to facilitate counterinsurgency efforts by overseeing and providing communication between all the PMCs operational in the country and US armed forces. Under the Reconstruction Operations Center, "Project Matrix" tracked and coordinated PMC activities. Transponders were issued to each contractor in order to facilitate more effective liaison between contractors and the military, and reporting back to central command any emergencies and insurgent activity. In this way, "Project Matrix" also had an intelligence-gathering function. Armed contractors visited construction sites collecting information on insurgent activity. According to Kristi Clemens, Aegis's executive vice-president, "Their mission is to provide 'ground truth' to the Army Corps."[126] They have been instrumental in US counterinsurgency efforts as tools of US statecraft to further US objectives in the region. PMCs were essential to the overall stabilization efforts. In many cases, the construction companies themselves hired their own security forces: "Since US forces were not available to protect those doing reconstruction work, such firms had no choice but to turn to PMCs to protect employees."[127]

In 2005–06, US advisors in Anbar province harnessed the defection of various Sunni tribes from the insurgent side and incorporated them into the US-led counterinsurgency effort.[128] US advisors capitalized on the defection of local Sunni tribes from the insurgency and enlisted them into a counterinsurgent program titled "Concerned Local Citizens," which was later given the name "Sons of Iraq."[129] This program trained and armed defected members of the insurgency, who were now handed a "special mandate to suppress, arrest, or kill local jihadist cadres."[130] The civilian defense force scheme formed a central component of US counter-insurgency strategy in Iraq and at its height coincided with the US troop "surge" in 2007. These forces provided intelligence to US personnel, ran security checkpoints in their areas, and conducted limited military operations against suspected insurgents and members of Al-Qaeda.[131] Initially concentrated primarily in the Anbar, Salah-ad-Din, Diyala, and Baghdad districts, the program rapidly expanded and counted on over 100,000 paramilitary fighters at its peak in 2007–08 in more than two-thirds of Iraqi provinces. These groups functioned as paramilitary security contractors for the US military, each "Son" receiving a $300 monthly salary for their services, the US allocating around $16 million a month for those payments in 2008.[132] US military leaders found the outsourcing of security and counterinsurgency functions to such militia forces extremely advantageous as they "realized that buying the loyalty of these tribes was cheaper and more effective than fighting them."[133]

During and after the US military withdrawal in 2011, the US-supported Iraqi government continued to work with various para-institutional forces. While much of the Iraqi armed forces contracted PMCs in advisory roles, training, and for specialist equipment, the Iraqi government tolerated and covertly supported various Shi'ite militia groups towards stabilization goals. Human rights organizations and other observers recorded that these militias conducted extra-judicial executions, arbitrary detentions, and intimidation of suspected members of the insurgency and political and ideological dissidents.[134] More recently (2014–18), US and Iraqi strategies against ISIS in Iraq and along its northern borders with Syria have been predicated on the surrogate use of force and the amplification of military efforts via non-state and privatized means. Both the US and Iraq have depended heavily on collaborations with militias and mercenaries outside the conventional military chains of command.[135] For its part, the Iraqi government encouraged and liaised directly with various Shi'ite and Sunni militias in combat against their common foes, with the Iraqi military coordinating both Shi'ite and Sunni militias. The US Air Force provided direct air support to Shi'ite and Kurdish militias to strengthen their efforts against ISIS.[136] US and international mercenaries and "volunteers" have joined some of these groups in their fight against ISIS. PMCs have also remained a significant component of US-led programs in Iraq and have aided in the continued presence of small contingencies of US soldiers and trainers in various support, training, and, occasionally, limited combat roles.[137]

Similar strategies of US statecraft have been applied across different parts of the world. Washington has directly and indirectly supported similar complexes of para-institutional agents in Yemen contributing to the stabilization of successive Yemeni governments against Houthi rebels and towards defeating Islamist forces (mainly Al-Qaeda in Yemen). Since 2012, the US has been increasingly involved in Yemen's civil war, training and assisting the Yemeni official Armed Forces in counterinsurgency.[138] The Yemeni government, in turn, has outsourced much of this fighting to local pro-government militias called "popular committees," paying many of these committees a wage for their security services and military collaboration.[139] After the Houthi takeover of the capital Sana'a in September 2014, a US-allied Saudi Arabian-led coalition increased interventionist activities in 2015, mainly relying on US-based (and therefore US-sanctioned) PMCs, mercenaries, and militias to drive back the insurgents and Islamist forces.[140] US Special Forces and CIA personnel have increasingly overseen some of this action, trained local irregular fighting forces, and directly participated with boots-on-the-ground operations and drone strikes.[141] US President Trump has made moves to deepen US proxy engagement in 2017 into the future.[142] US and Saudi involvement has hinged substantially on protecting oil pipelines that flow from Saudi Arabia and through Yemen to the Red Sea.

In Pakistan, alongside substantial US military assistance and training to Pakistan's official forces, US military leaders praised Pakistan's partnerships with "tribal militias" (Lashkars) comparing them to the "Sons of Iraq" program.[143] While there is little evidence to suggest that the US has supported these militia and paramilitary factions directly, the Pakistani government has. These militias form a crucial part of US-Pakistan strategy, overlapping with US and PMC activities in the Federally Administered Tribal Areas (FATA) of the Pakistan-Afghan border to counter the Taliban insurgency in an attempt to regain control of Afghanistan and to annihilate militant Jihadists. More directly, the CIA has hired PMCs, such as Blackwater, and backed militia forces from the Afghan side of the border to operate within Pakistan to furnish intelligence that would lead to the capture or assassination of "destabilizing" forces in Pakistan.[144] These forces have been instrumental to gathering information for US aerial drone strikes within Pakistan. Overall, US strategies towards Pakistan have developed a complex of military-paramilitary relationships. The US has largely sought to avoid directly committing US forces to combat missions in Pakistan and instead has aided in the development of a Pakistani-paramilitary nexus and relied on para-extended means of coercive statecraft.

In similar arrangements, US-backed insurgents, "freedom fighters," and mercenaries have served as the coercive spearhead of American unconventional warfare. These collaborations have followed strategic patterns whereupon local groups operate the front lines, capture and kill high valued targets on US "black lists" (called the "Disposition Matrix"), and mark targets for US aerial strikes. For example, US foreign policy towards Syria has followed a low-risk and low-visibility strategy in which the CIA and the DoD have supported local rebel and militia factions towards the twin objectives of toppling the Assad regime and fighting Islamic State (ISIS). The US had aimed for regime change in Syria and supported oppositional political movements long before the Arab Spring of 2011.[145] Once the multi-sided civil war broke out, the CIA and other private benefactors coordinated with and armed insurgent forces, such as the Free Syrian Army (FSA) towards the overthrow of the Assad regime.[146] In 2012, Obama reportedly turned down CIA director David Petraeus's proposals to increase US direct financing, training, and support to Syrian rebel forces.[147] However, in 2013, a US-led multinational consortium (with contributions from the UK, France, Saudi Arabia, Turkey, and Jordan among other countries primarily through the CIA) stepped up assistance to anti-Assad rebel forces which culminated in paramilitary training programs in neighboring countries and the provision of non-lethal aid.[148] The CIA, operating from Military Operations Command in Turkey and in Jordan continued to support selected militias with anti-tank missiles, armored cars, and other military hardware alongside training and intelligence support towards weakening the Assad regime.[149] According to emails released by Wikileaks, US PMCs aided in the training and operated alongside these Syrian oppositional forces against the Assad government in efforts to promote regime change in the country.[150]

As the Syrian crisis unfolded, US (and international) priorities shifted away from regime change towards fighting ISIS, and localized forces were recruited with the goal of destroying ISIS. In 2014, the Obama administration received Congressional approval for an overt "train and equip" program specifically set up, according to US military records, "to combat the Islamic State and other terrorist organizations in Syria and setting the conditions for a negotiated settlement to Syria's civil war."[151] Training, arms, and military and intelligence cooperation was provided to the Syrian Democratic Forces (SDF), a Kurdish-majority alliance of various militias and rebel factions, towards fighting ISIS. Special Forces advisors were operational in Syria to oversee these efforts and in some cases operate alongside their "irregular" Kurdish counterparts.[152] As part of this the US has also lent support to the Kurdish YPG (*Yekîneyên Parastina Gel*—People's Protection Units) in northern Syria.[153] Consistent with the US's distancing acts, these forces have been supported by various PMCs conducting intelligence and other similar support operations.[154] American air strikes have strengthened these forces and their fight against ISIS and other jihadist forces, such as Jhabat Al-Nusra.[155] The US Air Force has continued to provide air-strike support to the advances of US-backed SDF in subsequent offensives against ISIS. Further contributing to these efforts, mercenaries from various countries, including the US and UK have joined the fray fighting ISIS.[156]

US collaboration with "irregular" forces in Syria was beset by many issues. First, the complex of various groups the US sought to support to overthrow the Assad regime and to destroy ISIS were faced with inter- and intra-group rivalries. This posed serious issues for united efforts towards both objectives of removing Assad and fighting ISIS. Second, some groups switched sides or allegiances, and others worked closely with more "radical" groups the US and its allies were not willing to support publicly. Subsequently, the US faced charges that it was supporting international "terrorism" and the "Stop Arming Terrorists Act" was introduced in Congress.[157] The international environment— with Russian support for Assad and the Syrian government and armed forces, Turkey's displeasure with the US supporting a rebel Kurdish organization with ties to the PKK, and Iranian involvement in the region—have muted US enthusiasm for mounting an insurgency. Consequently, the US and its allies focused local collaborations and efforts on driving out ISIS and by September 2017, the US-backed SDF and US Special Forces had quashed the last of ISIS in Syria.[158]

In Libya, the US were quick to secretly support Libyan insurgents against Gaddafi. The US had long-standing interests in removing Gaddafi from power, which included unfettered access to oil resources, ideological disagreements, and security concerns, including Gaddafi's support for anti-Western forces.[159] The CIA had previously plotted to overthrow the regime during the Reagan years.[160] In the context of the Arab Spring movements that rattled the political establishment of the old guard across the Middle East, the CIA had their chance. Under secret orders authorized by Obama, CIA officers were dispatched to meet and help arm insurgent forces in Libya.[161] The US spent over $1 billion overall and led an international consortium, including Qatar, the UK, France, and others to support the rebels.[162] PMCs and private ex-service personnel were also contracted to help train the insurgents.[163] US and NATO air support and US-led air strikes were central to Libyan rebel victories. Many observers pointed to how the US-led NATO intervention overstepped its UN mandate UNSCR 1973 to protect civilians and became central to the overthrow of the Gaddafi regime.[164]

While the insurgency was ultimately victorious, capturing and killing the Libyan leader, the political aftermath has been unstable with a collection of competing factions and militias jockeying for power. Much of US strategy to counter the various militias has been to arm and train its own, and has focused on building alliances of local militias and political leaders with the ultimate goal of constructing a Libya congenial to US and transnational interests, which would safeguard against Islamist political and military advances.[165] The US and allied states have led the way in supporting militias and warlords to fight the ISIS groups that appeared in the power vacuum following Gaddafi's demise.[166] US military advisors planned to train US "friendly" militia forces in the Canary Islands as part of this strategy, but did not complete this mission.[167] The US-supported

Libyan GNA (Government of National Accord) cooperates with a loose collection of powerful militia formations and warlords. The Petroleum Facilities Guard militia, recently aligned with the GNA, served as another para-extensional security vanguard that indirectly defended US interests by ensuring the lucrative and strategic Libyan oil fields did not fall into the "wrong hands."[168]

Similar para-institutional complexes have developed as part of US statecraft elsewhere. In Somalia, the US has aligned itself with various militia factions to push back Jihadists and Al-Shabaab in 2003 onwards.[169] The CIA and Special Forces paid substantial sums of money to some groups for the capture and rendering of people suspected of supporting fundamentalists under a 2006 program called the "Alliance for the Restoration of Peace and Counterterrorism."[170] A US intelligence official confirmed the US was supporting a "collection of warlords" that helped furnish information for US aerial strike targets.[171] Evidence has also pointed to private companies running military operations in Somalia.[172] According to reports, more recently, in 2016, in conjunction with working alongside AMISOM (the African Union Mission to Somalia), the official Somali National Army, and a multitude of contractors, the US has continued to assist militias and warlord factions against Islamist militants.[173] In Ukraine, the US has supported the new government in its attempts to consolidate rule and counter Russian military advances in the country. The Ukrainian government has outsourced to semi-official and informal militias and foreign mercenaries in its war against pro-Russian insurgents.[174] US and private American civilian volunteers have allegedly supported some of these paramilitary groups, including the Azov Battalion.[175] PMCs have also buttressed Ukrainian efforts.

These examples help to highlight the scope of a US military-para-institutional nexus in the prosecution of stability operations around the world. US counterinsurgent and unconventional warfare statecraft has been exercised alongside composite and disparate public-private relations. This has involved the US working "with and through" a collection of "irregular" assets, in accordance with the "irregular warfare" doctrine, towards the combating of the growing prominence of Jihadi Salafists, the destabilization of unwanted regimes, and the stabilization of others. This has included working with militias, insurgent guerrillas, mercenaries, PMCs, and a host of other similar non-state armed groups. In the "War on Terror" and beyond, the US presided over a vast network of para-institutional agents to augment its influence and achieve its politico-military objectives. Some of these examples also help to highlight the flexible and contextual formation of these relationships. Rather than a "top-down" application of US statecraft policies, local context and agency has influenced how these arrangements form and the patterns of violence they employ. Para-institutional complexes in US statecraft arise and evolve through composite relations with various forces, groups, and individuals. There is an intersection of disparate forces working towards US stabilization processes. MNCs have also been key to these arrangements, liaising with US forces, local states, and para-institutional forces for the security of their businesses but also for coordination between these forces towards counterinsurgent goals. An invisible boundary is breached from security measures to proactive involvement in counterinsurgency.

Conclusion: The Admission of Essence

Global/local para-institutional forces now represent the principal coercive agents of US imperialism. Militias, mercenaries, and contractors play an increasingly central role in US statecraft abroad and the maintenance of a US-led liberalized global order. Such forces are now integral to US global military presence and its power projection around the world. In order to sustain extended US military relations and engagements, policymakers in Washington have had to rely on parallel military structures. PMCs, international mercenaries and local irregular and paramilitary units fill in the gaps where required and sometimes undertake the sensitive roles US "regular" armed forces are not willing or able to perform. The simultaneous US military engagements in Iraq and Afghanistan and substantial commitments in other parts of the world would not have been possible if it were left only to the US's "official" military and those of allied states. These arrangements have also supplied a politically expedient instrument to diminish military demands on US populations, replacing, for example, US soldier funerals with less media-sensitive contractors' body bags.[176] Yet, this was not a new feature of US imperialism. Patterns of contemporary US coercive statecraft have evolved and developed out of Cold War practices. "Talk of the 'Salvador Option,'" as Grandin says, "is not an indication of the failure of Washington's imperial policy but an admission of its essence."[177]

Conclusions

This book has examined the nature of US imperialism throughout the post-World War II era and traced the development of US outsourcing and relationships with a diverse range of para-institutional armed forces. Drawing from existing analyses of US foreign policy and its relation to global capitalism, it has argued that the objectives of US imperialism and the role of the US state in advancing global capital have been fairly consistent since the ascendancy of US power. The US has positioned itself as the ultimate guarantor for the stabilization of capitalist socio-economic relations and corresponding sets of political and economic policies, principally in resource rich areas of the global South. This has been pursued according to a "dual logic" whereupon both national and transnational capital interests have informed or driven US imperialism. While there are many means through which the US has exerted its influence and power, this book has focused on the military dimensions of US imperialism and in particular on counter-insurgency and unconventional warfare. These highly interventionist forms of coercive statecraft have been designed to aid in the policing of peripheral states towards the stabilization of beneficial political and economic arrangements abroad. US counterinsurgent interventions and military assistance to allied regimes have ultimately aimed at the containment of counter-hegemonic forces. Unconventional warfare, on the other hand, has been applied to undermine and overthrow regimes that threaten to follow alternative political developmental pathways outside of the US-led liberalized global order.

Within this context of US imperialism, this book has concentrated on how and why militias, mercenaries and contractors support US statecraft. It has traced the development of a nexus in the conduct of US statecraft between state military and para-institutional or non-state armed forces coordinated towards stabilization goals. Throughout the post-war period, US strategists have increasingly delegated central military tasks to, and formed public-private partnerships with a diverse range of para-institutional armed forces in order to stabilize state formations abroad which are conducive to global capital. Militias, paramilitary organizations, insurgents, "freedom fighters," mercenaries, PMCs, and many other forces have formed central components in US-supported counterinsurgency and unconventional warfare. This has occurred across different planes and levels of US intervention. On an international level, PMCs and other "private" actors have formed a para-extensional bedrock upon which the US has projected its global military presence. PMCs, for example, have increasingly provided the cushioning needed to uphold US military presence in locations across the world, undertaking central support roles for US armed forces and military. Within the local settings of US statecraft, forces such as militias and paramilitary forces as well as international actors, such as mercenaries and PMCs, have formed the basis of para-institutional complexes of intervention.

Crucially, this nexus has not been a standard package or set of policies, but represents patterns of outsourcing, co-optation and acquiescence that have emerged within local contexts. Rather than merely a "top-down" process of "outsourcing" or "privatization" (although these hierarchical delegations of force are also important), US planners have formed significant military-paramilitary alliances and flexible relations with diverse para-institutional actors. This has included harnessing transnational movements and leveraging local collaborations and paramilitary formations. Local paramilitary forces often autonomously develop within corresponding sets of interests and processes of capital accumulation as well as within dynamics during the course of counterinsurgent or unconventional conflict. US planners have harnessed local agency and power structures or facilitated it to contribute to the overall processes in US-led statecraft. The notion of a military-paramilitary nexus captures the intersection or interplay of coercive forces between the US military and various para-institutional actors and the ways in which they have come together or form a collective synergy towards stabilization goals. US planners have sought to coordinate these relations either directly or indirectly and have managed to do so to various degrees. In many cases, this has entailed a direct orchestration of para-institutional forces. In other cases, this has involved an indirect management and acquiescence to such groups.

In providing an overview of how and why militias, mercenaries, and contractors support US statecraft, this book has drawn attention to how a military-paramilitary nexus is embedded in the architecture of US imperialism. The US's ability to influence the politics and economics of peripheral states, principally in the global South, has relied to an increasing extent on outsourcing, collaborations with local agents, or more broadly, a para-extensional means of power projection. Significant public-private partnerships and state-paramilitary alliances have meant that a diverse range of para-institutional actors engage in order-making and the governance of peripheral states. While the size and scope of the US military and its expansive budget is truly impressive, and plays important roles in the international arena, the US's ability to coordinate and facilitate collective synergies of efforts towards the common enforcement of broader political and economic conditions abroad conducive to global capital has been much more important in forging a semblance of a US-led liberalized global order. Through these practices of coercive statecraft, US power has served to armor specific globalized political economies and helped to determine the developmental pathways of entire countries. This has had a profound effect on processes of state formation in the global South. The use and empowerment of certain para-institutional forces during US interventions and supported counterinsurgencies has had a substantial constitutive effect on the structure and organization of power in states in the South.[1] The use of pro-government militias, paramilitary forces, civilian defense forces, mercenaries, and PMCs also has effects on dynamics of violence and conflict in the course of these counterinsurgent or unconventional wars.[2] The impact on human rights violations, types of violence, and patterns of conflict, as well as on "which side wins," and the political and economic order of target states cannot be overstated.

Moreover, this analysis has highlighted the organization and articulation of power in a way that questions or transcends conventional notions of clearly distinct and separate "public" and "private" realms. The boundaries between "public" and "private" have often proven to be porous, perhaps better described as occupying a spectrum rather than a dichotomy. US statecraft has often relied on a revolving door between the "public" and "private" spheres. During the Cold War, this was often manifest in "sheep-dipping" to make US personnel look like "private" citizens on paper if captured. The US has also contracted businesses, such as Civil Air Transport and Air America, among many other semi-private companies (partially owned by the CIA). This has continued in various forms, such as in the concretization of PMCs and former US military personnel selling back their expertise to the US and approved recipients of US assistance. US statecraft has also consisted in the bringing together of various types of unofficial para-institutional or "irregular" forces as integral parts of the defense of particular orders. Diverse sets of coercive actors have been connected to and are a part of broader processes in spearheading capitalist accumulation and the stabilization of corresponding sets of socio-economic relations and political and economic arrangements.

The principal reasons for the formation of military-paramilitary nexus and for its development over time relate primarily to the nature of an informal US imperialism and its role within the advancement of global capitalism. First, the US has never been able or willing to engage in extensive and protracted militarized interventions to police the periphery, even where it has been deemed necessary. The managerial role the US has played in the international system has always been mediated to some extent through associated forces, whether through military assistance and training programs to allied states, or by building coalitions in US-led interventions, such as the "More Flags" program in Vietnam, or leveraging para-institutional mechanisms to aid in the burdens of the military dimensions of stabilization statecraft. Yet these stabilization responsibilities in its commitment to an Open Door strategy and managerial role have expanded significantly with the advancement of global capital. The growth of increasingly transnational and interconnected global markets and the industrialization of major economies, such as China's, have compounded global stabilization requirements. Transnational demand for mechanisms to "open up" noncompliant zones, to impose stability and order, and to repress counter-hegemonic dissent has strengthened alongside this growth.[3] US policymakers and military planners, in turn, have found ways to amalgamate various para-institutional means in US-led statecraft and interventions towards this US managerial role. Without these para-institutional connections and public-private partnerships, the US would be unable to sustain its global military presence and to extend coercive reach into distant lands. The use of proxy parallel armed groups such as PMCs and paramilitaries help to serve as "force-multipliers."

In relation to this, secondly, US interventions and US-led stabilization programs have had significant political costs, which US policymakers have been eager to avoid. Domestic public consciousness and the capacity to accept the loss of American lives have placed constraints on successive administrations' ability to deploy US soldiers to distant conflicts. This became particularly acute following the Vietnam War in both anti-war movements and the so-called "Vietnam syndrome." Similar domestic political costs have continued to be manifest in anti-war protests and a contemporary "body bag syndrome" during the "War on Terror."[4] International reputational and political costs have had similar constraints on US military endeavors. US policymakers attentive to international perceptions of the US as a leader of the liberal order have often had to "sell" US-led military campaigns to other core liberal states. The optics of anti-communism and the "War on Terror" have served as frameworks to do this. While US leadership and its managerial role in the international system have been broadly accepted with other core states' incorporation into the US-led order, US policymakers have often found it difficult to raise international approval for militarized interventions. For example, the muted consent, and in many cases outright opposition, to the Bush invasion of Iraq by international partners highlighted the limitations of American unilateralism. These political costs of intervention are contributing factors towards the increasing para-institutionalization of US statecraft. Cultivating various relationships with and delegating tasks to para-institutional forces has served as a way for Washington to "distance" itself from the coercive actions taken on its behalf.[5]

Finally, a series of strategic rationales have provided a basis for the growth of a nexus between US-led intervention and para-institutional forces. PMCs, for example, are commonly understood to be cheaper and more efficient than the military itself and provide specific military and technical expertise. The logics of market competition and privatization that rose to prominence in the 1980s have impacted the military sector.[6] According to US officials, for example, the wars in Iraq and Afghanistan would not have been possible without contractors due to the prohibitive costs and strains on resources in sending large military contingents abroad.[7] Local "irregular" groups of various tropes also ostensibly offer distinct advantages in US-led statecraft. Policymakers often highlight that paramilitary groups, for instance, offer cheap, fast, and expendable means of exerting force. Civilian militias, or civilian defense forces often reside within or near the area of operation, providing local knowledge of the people, language, and terrain, and thereby help solve the "identification problem." In addition, US military planners understand militias and other groups to have political functions.[8] They can, according to US military doctrines, help gain the active participation of local populations via "counter-organization," play a central role in local rule and state building, and can confer legitimacy to their governments. Finally, para-extensional forms of engagement often provide the plausible deniability policymakers desire and/or aid in distancing the US from interventionist charges, reducing the political costs of intervention and limiting scrutiny and accountability.[9]

These sets of practices that comprise this nexus have a longer historical genealogy in US foreign policy than is commonly recognized. They also have their roots in practices of US imperial statecraft. "Irregular warfare" and associated strategies such as an "indirect approach" and "surrogate warfare," for example, which have seen the US leverage local collaboration in the form of militias, paramilitaries, and other similar forces in many countries across the Middle East, have extended histories in US statecraft. US and US-supported counterinsurgency and unconventional warfare have often depended on composite relations with para-institutional forces. Privatization to PMCs in US foreign policy has also had a long history in US statecraft. US agencies had hired private and semi-private air-contractors for logistical capabilities for US covert operations, for combat operations, and as part of its assistance to allied countries throughout the Cold War, long before the global spread of private security industry in the 1990s and early 2000s. The end of the Vietnam War and the rise of neoliberal ideologies began to cement privatization into US foreign policy during the 1980s, a time which saw the creation of a diverse range of PMC providers. The US reliance on PMCs expanded significantly throughout the 1990s and into the "War on Terror." The early precursors to contemporary practices have their origins in covert operations in US statecraft during the Cold War. US outsourcing, privatization, and alliances with various paramilitary forces are much older practices than we often tend to think, and have developed within practices of US statecraft.

Practices of outsourcing, public-private partnerships and alliances with militias, mercenaries, and other similar forces are set to continue as prominent features in US foreign policy. The managerial role that the US occupies and the ways in which the US has served as a "stabilizer of last resort" for global capital, transnational interests, and those interests of other core capitalist states are set to remain intact. As long as there is a transnational (on behalf of transnational class) and international (on behalf of other core capitalist states) demand for stability in peripheral zones for the relatively fluid flow of resources and capital and protection for assets and investments against inimical calls for reform, the US is set to continue as the primary guarantor and enforcer of the current international system and the US-led liberalized global order. In this position, the US is likely to continue to pursue strategies which shift a good portion of these responsibilities, and the "dirty war" militaristic components in particular, to para-institutional forces. These practices have long rested on assorted strategic rationales within the broader schema of US informal imperialism. Such practices afford the US the ability to intervene in extreme circumstances and participate in the restructuring and reorientation of states in the global South towards global capital, without direct control or administration or even, in many cases, direct military intervention. Local collaboration is a fundamental feature of US imperial relations and the processes involved in the liberalization of political economies and the stabilization of their corresponding political and economic arrangements.

The Trump administration has promised a less interventionist, "disengaged" foreign policy in favor of a domestic focus on "making America great again." Trump proclaimed that "We will stop racing to topple foreign regimes that we know nothing about, that we shouldn't be involved with" and pledged, instead, as part of his foreign policy agenda, that he would focus on "destroying" ISIS.[10] However, while this signals that the US might not return to the unilateral interventionist impulse that the world witnessed under the George W. Bush presidency, it does not mean the US will become less involved in stabilization programs abroad. The logics of US statecraft constitute powerful drivers in US foreign policy to exert power through para-extensional means. Albeit with using a very different rhetoric and tone, Obama made similar promises during his presidency, only to advance these practices and ancillary modes of warfare that distanced the US from operations taken under its behalf. Trump is not Obama and this does not guarantee their continuation, but the structures and organization of power in the international arena and the US's military managerial role in global capital have cut across successive presidencies and it is unlikely that Trump will fundamentally alter the US's privileged position and shirk the responsibilities that come with it. In addition, Trump has already shown signs that he will continue to operate along similar lines as his predecessors. He has increased US military spending significantly and has already reversed his promises to withdraw from Afghanistan, vowing to send more troops in what has turned out to be the longest war in US history. More tellingly, alongside the discovery of rich mineral deposits and with the continuation of threats to the desired stability from the Taliban and other Islamist insurgents, policymakers in the Trump administration have unveiled proposals to create "new" paramilitary militia forces in Afghanistan and have contemplated proposals to hire PMCs to do much of the work previously reserved for US soldiers.[11] Trump has also authorized arming, training, and coordinating a collection of militia organizations in Syria. This was a central component of his plans to "destroy" ISIS in the region. Trump's promises for a less engaged foreign policy will most likely eschew direct interventions in favor of indirect methods, including outsourcing, privatization, and alliances with various militias and paramilitary formations as previous administrations have done. Outsourcing, privatization, and support for militias and other para-institutional forces are part of the essence of the American imperial project and the expansion of capital.

Notes

Introduction

1. Steven Hurst, *The United States and Iraq since 1979: Hegemony, Oil, and War* (Edinburgh: Edinburgh University Press, 2009), 129–131; Bob Woodward, "CIA Led Way with Cash Handouts," *Washington Post*, November 18, 2002.
2. Dana Priest and Josh White, "Before the War, the CIA Reportedly Trained a Team of Iraqis to Aid US," *Washington Post*, August 3, 2005; Mike Tucker and Charles Faddis, *Operation Hotel California* (Guilford: The Globe Pequat Press, 2009), 33–38.
3. Ewen Macaskill and Ian Traynor, "Bush Approves $92m to Train Iraqi Militia to Take On Saddam," *Guardian*, December 11, 2002.
4. Jim Garamone, "US Army Trains Free Iraqi Forces in Hungary," *American Forces Press Service*, US Department of State, February 24, 2003.
5. Greg Grandin, *Empire's Workshop: Latin America, the United States, and the Rise of the New Imperialism* (New York: Owl Books, 2006); Jeremy Scahill, *Blackwater: The Rise of the World's Most Powerful Mercenary Army* (New York: Nation Books, 2007), 349 and 377; Peter Maass, "The Way of the Commandos," *New York Times Magazine*, May 1, 2005.
6. Govinda Clayton and Andrew Thomson, "The Enemy of My Enemy is My Friend … the Dynamics of Self Defense Forces in Irregular War: The Case of the Sons of Iraq," *Studies in Conflict & Terrorism* 37, no. 11 (2014), 920–935; Stephen Biddle, Jeffrey Friedman, and Jacob Shapiro, "Testing the Surge: Why did Violence Decline in Iraq in 2007?" *International Security* 37, no. 10 (2012), 7–40; Daniel Green, "The Fallujah Awakening: A Case Study in Counterinsurgency," *Small Wars and Insurgencies*, 21, no. 4 (2010); Andrew Koloski and John Kolasheki, "Thickening the Lines: Sons of Iraq, A Combat Multiplier," *Military Review* (January 2009), 41–53.
7. See, for example, Shane Bauer, "Iraq's New Death Squad," *The Nation*, 6 (June, 2009).
8. The situation was much more nuanced than can adequately be portrayed here. Many observers worried that Shi'ite pro-government militias would fan the flames of sectarian conflict in the country as they targeted, with impunity, Sunni populations (the insurgency was largely Sunni based). US forces had a checkered past with Shi'ite militias. On the one hand, they often challenged the authority of the newly formed Iraqi state and had to be reined in. They also, at least early on in the conflict, challenged US presence and were later supported by Iran. On the other hand, they helped a fledging Iraqi military contain the Sunni insurgency. In other words, from roughly 2010–13, the US and the Iraqi government were

balancing the advantages and disadvantages of Shia militias. See Human Rights Watch, *Iraq: Pro-Government Militias' Trail of Death*, Human Rights Watch, July 31, 2014. www.hrw.org/news/2014/07/31/iraq-pro-government-militias-trail-death; Amnesty International, Iraq, *Absolute Impunity: Militia Rule in Iraq* (London: Amnesty International), October 14 , 2014. www.amnesty.org/en/documents/MDE14/015/2014/en/; Jack Healy and Michael Schmidt, "Political Role for Militants Worsens Fault Lines in Iraq," *New York Times*, January 5, 2012.

9. Molly Hennessey-Fiske and W. J. Hennigan, "The U.S. is helping to train Iraqi militias historically tied to Iran," *Los Angeles Times*, December 14, 2016; Tom O'Connor, "U.S. Military Set to Make $300 Million Deal to Arm Kurds Fighting ISIS in Iraq," *Newsweek*, April 20, 2017.

10. Christopher Doran, *Making the World Safe For Capitalism: How Iraq Threatened the US Economic Empire and Had to Be Destroyed* (London: Pluto Press, 2012); Eric Herring and Glen Rangwala, *Iraq in Fragments: The Occupation and its Legacy* (London: C. Hurst and Co., 2006); Doug Stokes, "Blood for Oil? Global Capital, Counter-Insurgency and the Dual Logic of American Energy Security," *Review of International Studies*, 33, no. 2 (2007).

11. Rosie Gray, "Erik Prince's Plan to Privatize the War in Afghanistan," *The Atlantic*, August 18, 2017. Also see Erik Prince, "Contractors, Not Troops, Will Save Afghanistan," *New York Times*, August 30, 2017.

12. See, for example, Representative John Terney, *Warlord, Inc.: Extortion and Corruption Along the U.S. Supply Chain in Afghanistan* (Washington, DC: Committee on Oversight and Government Reform, US House of Representatives, 2010); Committee on Armed Services United States Senate, *Inquiry into the Role and Oversight of Private Security Contractors in Afghanistan*. https://fas.org/irp/congress/2010_rpt/sasc-psc.pdf

13. These are all ways in which "surrogate warfare" or "irregular warfare" are described in doctrine. Please see: US Army, *Army Special Operations Forces: Unconventional Warfare: FM 3-05.130* (Washington, DC: Department of the Army, 2008); Isaac Peltier, "Surrogate Warfare: The Role of US Army SF," in R. Newton et al., eds., *Contemporary Security Challenges: Irregular Warfare and Indirect Approaches* (Hurlburt Field: JSOU Press, 2009), 55–85; Major Kelly Smith, *Surrogate Warfare for the 21st Century* (Ft. Leavenworth, KS: School of Advanced Military Studies, 2006).

14. See, for example, the following: Allison Stanger, *One Nation Under Contract: The Outsourcing of American Power and the Future of Foreign Policy* (London: Yale University Press, 2009); Tony Geraghty, *Soldiers of Fortune: A History of the Mercenary in Modern Warfare* (New York: Pegasus Books, 2009); Geraint Hughes, *My Enemy's Enemy: Proxy Warfare in International Politics* (Brighton: Sussex Academic Press, 2012).

15. Deborah Avant, *The Market for Force: The Consequences of Privatising Security* (Cambridge: Cambridge University Press, 2005); Laura A. Dickinson, *Outsourcing War and Peace* (New Haven, CT: Yale University Press, 2011), 3; Christopher Kinsey, *Corporate Soldiers and International Security: The Rise of Private Military Companies* (London: Routledge,

2007), 95; P. W. Singer, *Corporate Warriors* (Ithaca, NY: Cornell University Press, 2003), 49; Kinsey, *Corporate Soldiers and International Security*, 1.

16. Andreas Krieg, "Externalizing the Burden of War: The Obama Doctrine and US Foreign Policy in the Middle East," *International Affairs*, 92, no. 1 (2016), 97–113; on the "new wars," see Mary Kaldor, *New and Old Wars: Organised Violence in a Global Era* (New York: John Wiley & Sons, 2013); On fourth-generational warfare, see William S. Lind et al., "The Changing Face of War: Into the Fourth Generation," *Marine Corps Gazette*, October, 1989, 22–26.

17. Christopher Layne, *The Peace of Illusions: American Grand Strategy from 1940 to the Present* (Ithaca, NY: Cornell University Press, 2006); Simon Bromley, "The Logic of American Power in the International Capitalist Order," in Alejandro Colas and Richard Saull, eds., *The War on Terrorism and the American 'Empire' After the Cold War* (London: Routledge, 2006), 44–64; Leo Panitch and Sam Gindin, *Global Capitalism and American Empire* (London: Merlin Press, 2004); Ray Kiely, "United States Hegemony and Globalisation: What Role for Theories of Imperialism?" *Cambridge Review of International Affairs*, 19, no. 2 (2006), 205–221; Ray Kiely, *Empire in the Age of Globalization: US Hegemony and Neoliberal Disorder* (London: Pluto Press, 2005).

18. Doug Stokes and Sam Raphael, *Global Energy Security and American Hegemony* (Baltimore, MD: John Hopkins University Press, 2010); Doug Stokes, "The Heart of Empire?: Theorising US Empire in an Era of Trans-national Capitalism," *Third World Quarterly*, 26, no. 2 (2005), 217–236; Michael T. Klare and Peter Kornbluh, "The New Interventionism," in idem, eds., *Low Intensity Warfare: Counterinsurgency, Proinsurgency and Anti-Terrorism in the Eighties* (New York: Pantheon Books, 1988), 3–20.

19. Carl Boggs, "Introduction: Empire and Globalization," idem, ed., in *Masters of War: Militarism and Blowback in the Era of American Empire* (New York: Routledge, 2003); Michael Parenti, "The Logic of US Intervention," in Carl Boggs, ed., *Masters of War* (London: Routledge, 2003), 19–36; Michael Parenti, *Face of Imperialism* (London: Routledge, 2015).

20. Catherine Lutz and Cynthia Enloe, eds., *The Bases of Empire: The Global Struggle Against US Military Posts* (New York: New York University Press, 2009); David Vine, *Base Nation: How US Military Bases Abroad Harm America and the World* (New York: Metropolitan Books, 2015).

21. Barry R. Posen, "Command of the Commons: The Military Foundation of US Hegemony," *International Security*, 28, no. 1 (2003): 5–46.

22. Max Boot, *The Savage Wars of Peace: Small Wars and the Rise of American Power* (New York: Basic Books, 2003); Robert M. Cassidy, *Counterinsurgency and the Global War on Terror* (Stanford, CA: Stanford University Press, 2008); David Ucko, *The New Counterinsurgency Era: Transforming the US Military for Modern Wars* (Washington, DC: Georgetown University Press, 2009).

23. See, for example, Lesley Gill, *The School of the Americas: Military Training and Political Violence in the Americas* (Durham, NC: Duke University Press, 2004).

24. Paul Macdonald, *Networks of Domination: The Social Foundations of Peripheral Conquest in International Politics* (Oxford: Oxford University Press, 2014); Victor Gordon Kiernan, *Colonial Empires and Armies, 1815–1960* (Montreal: McGill-Queen's University Press, 1998); David Killingray and David E. Omissi, *Guardians of Empire: The Armed Forces of the Colonial Powers c. 1700–1964* (Manchester: Manchester University Press, 1999); Janice E. Thomson, *Mercenaries, Pirates, and Sovereigns: State-Building and Extraterritorial Violence in Early Modern Europe* (Princeton, NJ: Princeton University Press, 1996).

25. See, for instance, Chalmers Johnson, *The Sorrows of Empire* (London: Verso, 2004), 131; Alejandro Colas and Bryan Mabee, eds., *Mercenaries, Pirates, Bandits and Empires* (London: Hurst and Company, 2010).

26. See Kiernan, *Colonial Empires and Armies, 1815–1960*; Myron Echenburg, *Colonial Conscripts* (Portsmouth: Heinemann, 1991).

27. Guy Arnold, *Mercenaries: The Scourge of the Third World* (New York: St Martin's Press, 1999), 147–161; also for British use of mercenaries in the African continent during the 1970s, see Wilfred Burchett and D. Roebuck, *The Whores of War: Mercenaries Today* (New York: Penguin Books, 1977); see Kinsey, *Corporate Soldiers*.

28. Geraghty, *Soldiers of Fortune*, 79–89. Clive Jones, "Britain Covert Action and the Yemen Civil War, 1962–1967," in Zach Levey and Elie Podeh, eds., *Britain and the Middle East: From Imperial Power to Junior Partner* (Eastbourne: Sussex Academic Press, 2007), 253–257.

29. Hughes, *My Enemy's Enemy*, 6.

30. See also Robert Holden, *Armies Without Nations: Public Violence and State Formation in Central America 1821–1960* (Oxford: Oxford University Press, 2004), 14; Adam Jones, "Review: Parainstitutional Violence in Latin America; Violence in Colombia, 1999–2000," *Latin American Politics and Society* 46, no. 4 (Winter, 2004), 127–148; See also Martha K. Huggins, ed., *Vigilantism and the State in Modern Latin America: Essays on Extralegal Violence* (New York: Greenwood Press, 1991).

31. US Army, *Army Special Operations Forces: Unconventional Warfare: FM 3-05.130* (Washington, DC: Department of the Army, 2008), 1–3.

32. Armed Forces of the United States, *Joint Tactics, Techniques and Procedures for Foreign Internal Defense: Joint Publication 3-07.1* (Joint Chiefs of Staff, 2004), Glossary, p. 8; US Army, *Doctrine for Special Forces Operations: Field Manual 31-20* (Washington, DC: Department of the Army, 1990), Glossary, p. 10.

33. Patricia Owens, "Distinctions, Distinctions: 'Public' and 'Private' Force?" *International Affairs* 84, no. 5 (2008), 977–990; Tarak Barkawi, "State and Armed Force in International Context," in Alejandro Colas and Bryan Mabee, eds., *Mercenaries, Pirates, Bandits and Empires* (London: Hurst and Company, 2010), 33–54.

34. Rita Abrahamsen and Michael Williams, *Security Beyond the State: Private Security in International Politics* (Cambridge: Cambridge University Press, 2011); Rita Abrahamsen and Michael C. Williams, "Security Beyond the State: Global Security Assemblages in International Politics," *International*

Political Sociology, 3, no. 1 (2009), 1–17; Laleh Khalili, "Tangled Webs of Coercion: Parastatal Production of Violence in Abu Ghraib," in Laleh Khalili and Jillian Schwedler, eds., *Policing and Prisons in the Middle East: Formations of Coercion* (London: Hurst, 2010).

1. US Imperial Statecraft and Para-Institutional Forces

1. The original "Open Door" policy referred to a US policy proposal to various countries regarding the terms and conditions under which they could do business with China. It provided that all or each of the other countries should have equal access to Chinese ports for trade (rather than one state dominating trade relations with China as was customary in European colonialism). It also added that China should maintain sovereignty over its own territory and preside over these free trade arrangements. See B.A. Elleman, *International Competition in China, 1899–1991: The Rise, Fall, and Restoration of the Open Door Policy* (New York: Routledge, 2015).

   However, the term has been adapted and is often used to describe US imperial policies more generally, but more specifically referring to the strategy of maintaining foreign state's sovereignty (i.e. anti-colonialism) and preserving "open" borders for free trade and capital flows. See Christopher Layne, *The Peace of Illusions: American Grand Strategy from 1940 to the Present* (Ithaca, NY: Cornell University Press, 2006). The term "Open Door" policy was popularized by William Appleman Williams, *The Tragedy of American Diplomacy. Fiftieth Anniversary Edition* (New York: W.W. Norton and Company, 2009 [1959]).

2. Harvey, *New Imperialism*; Fred Halliday, "The Pertinence of Imperialism," in Mark Rupert and Hazel Smith, eds., *Historical Materialism and Globalization* (London: Routledge, 2002), 75–89; James Petras and H. Veltmeyer, *Globalization Unmasked: Imperialism in the 21st Century* (London: Zed Books, 2001); William I. Robinson, *A Theory of Global Capitalism: Production, Class and State in a Transnational World* (Baltimore, MD: John Hopkins University Press, 2004).

3. Leo Panitch and Sam Gindin, *Global Capitalism and American Empire* (London: Merlin Press, 2004), 24.

4. Laurence Shoup and William Minter, *Imperial Brain Trust: The Council on Foreign Relations and United States Foreign Policy* (New York: Authors Choice Press, 2004 [1977]), 117–177; Gabriel Kolko, *Confronting the Third World: United States Foreign Policy 1945–1980* (New York: Pantheon Books, 1988), and for example, Leo Panitch and Sam Gindin, *The Making of Global Capitalism: The Political Economy of American Empire* (London: Verso Books, 2012).

5. Michael Cox, *US Foreign Policy After the Cold War: Superpower without a Mission?* (London: RUSI, 1995), 5.

6. Andrew Bacevich, *American Empire* (London: Harvard University Press, 2002), 3.

7. Panitch and Gindin, *Global Capitalism and American Empire*; Leo Panitch and Sam Gindin, "The Unique American Empire," in Alejandro Colas and

Richard Saull, eds., *The War on Terrorism and the American 'Empire' After the Cold War* (London: Routledge, 2006), 24–43, at 42. See also Panitch and Gindin, *The Making of Global Capitalism*.

8. Simon Bromley, "The Logic of American Power in the International Capitalist Order," in Colas and Saull, eds., *The War on Terrorism*, 44–64.

9. National Security Council, "A Report to the National Security Council— NSC 68," April 12, 1950. President's Secretary's File, Truman Papers. Washington, DC: US National Security Council, 28.

10. US Office of the President, *A National Security Strategy for A New Century* (Washington, DC: The White House, 1997).

11. US Office of the President. *National Security Strategy* (Washington, DC: The White House, 2010).

12. See David S. Painter, "Explaining US Relations with the Third World," *Diplomatic History*, 19, no. 3 (Summer, 1995), 525–549; Kolko, *Confronting the Third World*; Williams, *The Tragedy of American Diplomacy*.

13. US Office of the President, *United States Overseas Internal Defense Policy* (Washington, DC: National Security Action Memorandum 182, 1962), 8.

14. Vladimir Lenin. *Imperialism: The Highest Stage of Capitalism* (Chippendale: Resistance Books, 1999).

15. Doug Stokes, *America's Other War: Terrorizing Colombia* (New York: Zed Books, 2005), 230. Also see Doug Stokes, "The Heart of Empire?: Theorising US Empire in an Era of Transnational Capitalism," *Third World Quarterly*, 26, no. 2 (2005), 217–236; Doug Stokes, "Blood for Oil? Global Capital, Counter-Insurgency and the Dual Logic of American Energy Security," *Review of International Studies*, 33, no. 2 (2007), 245.

16. Doug Stokes, "Blood for Oil?"; Robinson, *A Theory of Global Capitalism*, 140.

17. National Security Council, "National Security Problems Concerning Free World Petroleum Demands and Potential Supplies," Report to the National Security Council by the Departments of State, Defense, the Interior, and Justice: NSC 138/1, US Office of the Historian, *Foreign Relations of the United States 1952–1954, General: Economic and Political Matters*, Volume 1, part 2 (Washington, DC, January 5, 1953). https://history.state.gov/historicaldocuments/frus1952-54v01p2/d159

18. National Security Council, "Note by the Executive Secretary to the National Security Council on Long-Range U.S. Policy Toward the Near East" National Security Council Report, US Office of the Historian, *Foreign Relations of the United States, 1958–1960, Near East Region: Iraq, Iran, Arabian Peninsula*, Volume 7. https://history.state.gov/historicaldocuments/frus1958-60v12/d5

19. Robinson, *A Theory of Global Capitalism*, 130–140.

20. Ibid., 138; emphasis added.

21. Richard Saull. *The Cold War and After: Capitalism, Revolution and Superpower Politics* (London: Pluto Press, 2007), 53–54; Doug Stokes, "Why the End of the Cold War Doesn't Matter: The US War of Terror in Colombia," *Review of International Studies*, 29 (2003), 569–585; Richard

Saull, *Rethinking Theory and History in the Cold War: The State, Military Power, and Social Revolution* (London: Frank Cass, 2001).

22. Layne, *The Peace of Illusions*, 3.

23. David N. Gibbs, "Pretexts and US Foreign Policy: The War on Terrorism in Historical Perspective," in Joseph Peschek, ed., *The Politics of Empire: War Terror and Hegemony* (Abingdon: Routledge, 2006).

24. Bromley, *The Logic of American Power*, 44–64; Jim Glassman, "The New Imperialism? On Continuity and Change in US Foreign Policy," *Environment and Planning*, 37, no. 9 (2005), 1527–1544.

25. Ruth Blakeley, *State Terrorism and Neoliberalism: The North in the South* (New York: Routledge, 2009), 73.

26. Bromley, *The Logic of American Power*, 48; Robinson, *A Theory of Global Capitalism*; Tarak Barkawi, *Globalisation and War* (Oxford: Rowman and Littlefield Publishers, 2006).

27. Panitch and Gindin, *Global Capitalism and American Empire*, 39.

28. Kolko, *Confronting the Third World*, 11.

29. David S. Painter, "Explaining US Relations with the Third World," *Diplomatic History*, 19, no. 3 (Summer, 1995), 525–549, 534.

30. On different types of US power, see Bryan Mabee, *Understanding American Power: The Changing World of US Foreign Policy* (New York: Palgrave Macmillan, 2013).

31. David Ekbladh, *The Great American Mission: Modernization and the Construction of an American World Order* (Princeton, NJ: Princeton University Press, 2011); Michael E. Latham, *The Right Kind of Revolution: Modernization, Development, and US Foreign Policy from the Cold War to the Present* (Ithaca, NY: Cornell University Press, 2011).

32. David Harvey, "Neoliberalism as Creative Destruction," *The Annals of the American Academy of Political and Social Science* 610, no. 1 (2007), 21–44; Bill Cooke, "The Managing of the (Third) World," *Organization* 11, no. 5 (2004), 603–629; Michael Hudson, *Super Imperialism: The Origin and Fundamentals of World Domination* (London: Pluto Press, 2003).

33. Lars Schoultz, *Beneath the United States: A History of US Policy Toward Latin America* (London: Harvard University Press, 1998), 34–67.

34. Stokes and Raphael, *Global Energy Security and American Hegemony*, 56–58.

35. US Office of the President, *United States Overseas Internal Defense Policy*, 3.

36. US Office of the President "United States Policy on Internal Defense in Selected Foreign Countries" Paper Approved by the Senior Interdepartmental Group, US Office of the Historian, *Foreign Relations of the United States, 1964–1968*, Volume X, National Security Policy, Washington, DC, 1968. https://history.state.gov/historicaldocuments/frus1964-68v10/d204

37. US Executive Secretary, "United States Objectives and Policies with Respect to the Arab States and Israel," A Report to the National Security Council. Washington, DC, April 7, (1952). https://nsarchive2.gwu.edu// NSAEBB/NSAEBB78/propaganda%20059.pdf

38. William S. Cohen, *Secretary of Defense Report of the Quadrennial Defense Review* (Washington, DC: Department of Defense [1997]).

39. George W. Bush, *The National Security Strategy of the United States of America,* (Washington, DC: The White House, March 2006). http://nssarchive.us/NSSR/2006.pdf; Barak Obama, *National Security Strategy* (Washington, DC: The White House, 2015).

40. Nathan Freier et al., *At Our Own Risk: DoD Risk Assessment in a Post-Primacy World* (Washington, DC: Strategic Studies Institute and US Army War College Press, June 2017). https://ssi.armywarcollege.edu/pubs/display.cfm?pubID=1358

41. US Armed Forces, *Stability: Joint Publication 3-07* (Washington, DC: Joint Chiefs of Staff, August 3, 2016). www.dtic.mil/doctrine/new_pubs/jp3_07.pdf

42. Odd Arne Westad, *The Global Cold War* (Cambridge: Cambridge University Press, 2007), 111.

43. Michael J. Sullivan, *American Adventurism Abroad: 30 Invasions, Interventions and Regime Changes since WW II* (London: Praeger, 2004).

44. Thomas Friedman, *The Lexus and the Olive Tree* (New York: Anchor Books, 2000), 466.

45. Layne, *The Peace of Illusions*, 36.

46. Kolko, *Confronting the Third World*, 55.

47. Beau Grosscup, "The American Doctrine of Low Intensity Conflict in the New World Order," in A. Grammy and K. Bragg, eds., *United States Third World Relations in the New World Order* (New York: Nova Publishers, 1996), 58.

48. Stokes and Raphael, *Global Energy Security*; Michael McClintock, *Instruments of Statecraft: U.S. Guerrilla Warfare, Counterinsurgency, and Counterterrorism, 1940–1990* (New York: Pantheon Books, 1992); Michael McClintock, "American Doctrine and Counterinsurgent State Terror," in Alexander George, ed., *Western State Terrorism* (Oxford: Polity Press, 1991).

49. US Army, *Military Operations in Low Intensity Warfare: FM 100-20* (Washington, DC: Headquarters, Department of the Army, 1990), 1-1.

50. US Office of the President, "Memorandum of a Conference With the President, White House, Washington, DC, 29 March, 1957, 11:15 a.m." US Office of the Historian. *Foreign Relations of the United States 1955–1957, National Security Policy*, Volume XIX. https://history.state.gov/historicaldocuments/frus1955-57v19/d113

51. US Department of State, "Memorandum From the Deputy Assistant Secretary of State for Politico-Military Affairs (Kitchen) to the Counselor and Chairman of the Policy Planning Council (Rostow)/1/," *Foreign Relations, 1964–1968*, Volume X, National Security Policy, Washington, DC, March 12, 1964. https://2001-2009.state.gov/r/pa/ho/frus/johnsonlb/x/9016.htm

52. See, for example, Armed Forces of the United States, *Joint Tactics, Techniques and Procedures for Foreign Internal Defense: Joint Publication 3-07.1* Joint Chiefs of Staff, 2004.

53. US Air Force, *Foreign Internal Defense: Air Force Doctrine Document 2-3.1*, Headquarters, US Air Force, 15 September, 2007. https://fas.org/irp/doddir/usaf/afdd3-22.pdf; also see US Army. *Foreign Internal Defense: ATP 3-05.2*. Washington, DC Headquarters: Department of the Army, 2015. www.apd.army.mil/epubs/DR_pubs/DR_a/pdf/web/atp3_05x2.pdf

54. This is an updated estimate based on previous estimates in the following reports: Lora Lumpe, "US Foreign Military Training: Global Reach, Global Power, and Oversight Issues," *Foreign Policy in Focus* (May, 2002), 26; Michael Parenti, "The Logic of US Intervention," in Carl Boggs, ed., *Masters of War* (London: Routledge, 2003), 19–36. According to these sources, since the end of World War II, the US has spent around $240 billion in training and equipping around 2.3 million members of foreign militaries from well over 80 countries worldwide. Adding updated reports and estimates from between Fiscal Year 2001 and Fiscal Year 2016 (the last year that figures are available for), the US has spent over $300 billion and trained many more members of foreign militaries. This does not include those trained by US PMCs and training programs contracted directly by the recipient country. The actual figures of US training will be much higher than these estimates. For other drawdowns of US military expenditures, see: US Department of Defense, *Foreign Military Training Fiscal Years 2015 and 2016*. (Washington, DC: US Department of Defense and US Department of State Joint Report to Congress, 2016). www.state.gov/documents/organization/265162.pdf, II-2; US Department of State, *International Military Education and Training (IMET)* (US Department of State Website. 2017). https://2009-2017.state.gov/t/pm/ppa/sat/c14562.htm (accessed November 17, 2017); US Special Operations Command, Operation and Maintenance, Defense-Wide Fiscal Year (FY) 2016 Budget Estimate. http://comptroller.defense.gov/Portals/45/Documents/defbudget/fy2016/budget_justification/pdfs/01_Operation_and_Maintenance/O_M_VOL_1_PART_1/USSOCOM_PB16.pdf

55. Parenti, *The Logic of US Intervention*, 20.

56. Ruth Blakeley, "Still Training to Torture? US Training of Military Forces from Latin America," *Third World Quarterly*, 27, no. 8 (2006), 1439–1461; Lesley Gill, *The School of the Americas: Military Training and Political Violence in the Americas* (Durham, NC: Duke University Press, 2004); Jack-Nelson Pallmeyer, *School of Assassins: Guns, Greed, and Globalization* (Maryknoll, NY: Orbis Books, 2001).

57. Lumpe, *US Foreign Military Training*, 1.

58. Stephanie Neuman, *Military Assistance in Recent Wars* (New York: Praeger, 1986), 28–29.

59. Stokes and Raphael, *Global Energy Security and American Hegemony*, 60.

60. Johnson, *The Sorrows of Empire*, 132.

61. US Government Accounting Office. *International Military Education and Training: Agencies Should Emphasize Human Rights Training and Improve Evaluations*, US Government Accountability Office, 2011; US Department of Defense, *Foreign Military Training Fiscal Years 2015 and 2016*. Washington, DC: US Department of Defense and US Department of State Joint Report

to Congress, 2016. www.state.gov/documents/organization/265162.pdf, II-2.; *Foreign Military Training Fiscal Years 2014 and 2015*. Washington, DC: US Department of Defense and US Department of State Joint Report to Congress, 2015. www.state.gov/documents/organization/243009.pdf; US Department of State, *International Military Education and Training (IMET)*. US Department of State Website, 2017. https://2009-2017.state.gov/t/pm/ppa/sat/c14562.htm (accessed November 17, 2017).

62. US Department of State, *International Military Education and Training (IMET)*. US Department of State Archive, 2009. https://2001-2009.state.gov/t/pm/65533.htm (accessed November 17, 2017).

63. John Rudy and Ivan Eland, "Special Operations Military Training Abroad and its Dangers," *Foreign Policy Briefing: CATO Institute*, 53 (June, 1999).

64. USSOCOM, *FY2013 Budget Highlights United States Special Operations Command* (Tampa Point, FL: USSOCOM, 2012); see also Andrew Feickert, *U.S. Special Operations Forces (SOF):Background and Issues for Congress* (Washington, DC: CRS, 2012); US Congress 2012, *National Defense Authorization Act for Fiscal Year 2013*, 112th Congress, 2nd Session https://theintercept.com/2016/09/08/documents-show-u-s-military-expands-reach-of-special-operations-programs/; Weisberger, Marcus, "Peeling the Onion Back on the Pentagon's Special Operations Budget", *Defense One*, January 27, 2015 www.defenseone.com/business/2015/01/peeling-onion-back-pentagons-special-operations-budget/103905/; US Special Operations Command , Operation and Maintenance, Defense-Wide Fiscal Year (FY) 2016 Budget Estimate http://comptroller.defense.gov/Portals/45/Documents/defbudget/fy2016/budget_justification/pdfs/01_Operation_and_Maintenance/O_M_VOL_1_PART_1/USSOCOM_PB16.pdf

65. Nina M. Serafino, *Security Assistance Reform: "Section 1206" Background and Issues for Congress* (Congressional Research Service, 2012).

66. US Department of State, *Foreign Military Training Fiscal Years 2013 and 2014: Joint Report to Congress*, Volume 1 (2015) www.state.gov/documents/organization/230192.pdf

67. Jochen Hippler, "Counterinsurgency and Political Control," *INEF Report* 81 (2006); James Kiras, "Irregular Warfare: Terrorism and Insurgency," in John Baylis, James J. Wirtz, and Colin Gray, eds., *Strategy in the Contemporary World* (Oxford: Oxford University Press, 2010, third ed.), 185–207, p. 187.

68. Hippler, *Counterinsurgency and Political Control*, 54.

69. Burnett and Whyte (2005), as quoted in Hippler, "Counterinsurgency and Political Control," 54.

70. Adam Moore, "US Military Logistics Outsourcing and the Everywhere of War," *Territory, Politics, Governance*, 5 no. 1 (2017), 5–27; Sean McFate, *The Modern Mercenary: Private Armies and What They Mean for World Order* (Oxford: Oxford University Press, 2014), 44; Bruce E. Stanley, *Outsourcing Security: Private Military Contractors and US Foreign Policy* (Lincoln, NE: University of Nebraska Press, 2015), 9.

71. See, for example, any of the following unconventional warfare manuals: CIA, *Power Moves Involved in the Overthrow of an Unfriendly Government*

(Langley, VA: CIA, 1970) www.cia.gov/library/readingroom/docs/DOC_0000919937.pdf; US Army, *Handbook for Aggressor Insurgent Warfare, FM 30-104* (Washington, DC: Department of the Army, September 1967); US Army, *Guerrilla Warfare and Special Forces Operations FM 31-21* (Washington, DC: Department of the Army, 1961); US Army, *Unconventional Warfare Mission Planning Guide for the Special Forces Operational Detachment—Alpha Level (TC 18-01.1).* (Washington, DC: Department of the Army, October 2016); US Army, *Unconventional Warfare Pocket Guide* (Fort Bragg, NC: US Army Special Operations Command, Deputy Chief of Staff G3, Sensitive Activities Division G3X, AOOP-SA, 2016).

72. US Army, *Army Special Operations Forces: Unconventional Warfare: FM 3-05.130* (Washington, DC: Department of the Army, 2008), 1-3.

73. A 1958 Special Forces training manual highlighted how US operatives were meant to mobilize insurgent or guerrilla armies to do their bidding: the "existence of organized guerrilla forces is not assumed … in many situations Special Forces teams will be required to establish contact with local inhabitants *to initiate the development* of friendly elements into effective guerrilla forces" (emphasis added). US Army, *Guerrilla Warfare and Special Forces Operations FM 31-21* (Washington, DC: Headquarters, Department of the Army, 1958), 19.

74. US unconventional warfare doctrine was clear that US policymakers and military commanders must examine the ideologies and political causes of insurgent groups with the ultimate objective to "organize, train, and further develop *existing and latent guerrilla potential* into guerrilla forces" (emphasis added). US Army, *Special Forces Operations: FM 31-21* (Washington, DC: Department of the Army, 1969), 20.

75. One example of an informal alliance includes US support for the Kuomintang in China. The Kuomintang were a nationalist political party and movement that, after the communist revolution and the formation of the People's Republic of China in 1949, became an oppositional guerrilla forces seeking to overthrow the newly established communist regime. The US aided the Kuomintang in pursuit of its own geopolitical and economic interests. See Richard Gibson and Wenhua Chen, *The Secret Army: Chiang Kai-Shek and the Drug Warlords of the Golden Triangle* (Hoboken, NJ: Wiley and Sons, 2011); Sarah-Jane Corke, *US Covert Operations and Cold War Strategy: Truman Secret Warfare and the CIA 1945–1953* (Abingdon: Routledge, 2008), 114–119.

76. US Army, *Counterguerrilla Operations: FM 31-16*, 40, and US Army, *US Army Counterinsurgency Forces: FM 31-22*, 19, 82; US Army, *Stability Operations: FM 31-23*, Section 8; US Army, *Tactics in Counterinsurgency: FM 3-24.2* (Washington, DC: Department of the Army, 2009), 3-22.

77. US Army, *Stability Operations: FM 3-07* (Washington, DC: Department of the Army, 2008). Also see US Army, *Threat Force Paramilitary and Nonmilitary Organizations and Tactics: TC 31-93.3* (Washington, DC: Department of the Army, 2003), 1-38.

78. David T. Mason and Dale A. Krane, "The Political Economy of Death Squads: Toward a Theory of the Impact of State-Sanctioned Terror,"

International Studies Quarterly, 33 (1989), 175–198; Miles Wolpin, "State Terrorism and Death Squads in the New World Order," in K. Rupesinghe and R. Marcial, eds., *The Culture of Violence* (New York: United Nations University Press, 1994). See also literature on pro-government militias in Colombia such as Jasmin Hristov, *Paramilitarism and Neoliberalism: Violent Systems of Capital Accumulation in Colombia and Beyond* (London: Pluto Press, 2014); Stokes, *America's Other War.*

79. Ariel Ahram, *Proxy Warriors* (Stanford, CA: Stanford University Press, 2011); Ariel Ahram, "Origins and Persistence of State-Sponsored Militias: Path Dependent Processes in Third World Military Development," *Journal of Strategic Studies*, 34, no. 4 (2011), 531–556.

80. Shane Joshua Barter, "State Proxy Or Security Dilemma? Understanding Anti-Rebel Militias in Civil War," *Asian Security*, 9, no. 2 (2013), 75–92.

81. Richard Jackson, *Writing the War on Terrorism: Language, Politics and Counter-terrorism* (Manchester: Manchester University Press, 2005); Richard Jackson, "The Ghosts of State Terror: Knowledge, Politics and Terrorism Studies," *Critical Studies on Terrorism*, 1, no. 3 (2008), 377–392. This can also help explain why most PMCs are broadly conducive to Western interests. PMCs don't offer their services to everyone equally. US-based PMCs need a license from the State Department to operate and so the government has administrative control over what they can and cannot do. However, PMCs around the world do not either, and are attached to and part of a US-led global order. Those PMCs that do, such as Malhama Tactical, are quickly condemned and made illegal.

2. Covert Regime Change in the Early Cold War: "Power Moves Involved in the Overthrow of an Unfriendly Government"

1. CIA, *Power Moves Involved in the Overthrow of an Unfriendly Government* (Washington, DC: CIA, 1970).

2. Stephen Kinzer, *Overthrow* (New York: Times Books, 2006); John Prados, *Presidents' Secret Wars* (Chicago, IL: Elephant Paperbacks, 1996), 199; William Blum, *Killing Hope: US Military and CIA Interventions since WW2* (London: Zed Books, 2002); William Blum, *Rogue State: A Guide to the World's Only Superpower* (London: Zed Books, 2003, second ed.); Michael J. Sullivan, *American Adventurism Abroad: 30 Invasions, Interventions and Regimes Changes since WW II* (London: Praeger, 2004).

3. John Prados, *Safe For Democracy: The Secret Wars of the CIA* (Chicago, IL: Ivan Dee, 2006), 199.

4. Christopher Robbins, *Air America* (New York: Avon Books, 1979 [1990]), xii.

5. Greg Grandin, *Empire's Workshop* (New York: Owl Books, 2006), 111.

6. Michael McClintock, *Instruments of Statecraft: U.S. Guerrilla Warfare, Counterinsurgency, and Counterterrorism, 1940–1990* (New York: Pantheon Books, 1992); Kinzer, *Overthrow.*

7. CIA, *Power Moves Involved in the Overthrow of an Unfriendly Government* (1970).

8. US Army, *Guerrilla Warfare and Special Forces Operations FM 31-21* (Washington, DC: Department of the Army, 1958); US Army, *Guerrilla Warfare and Special Forces Operations FM 31-21* (Washington, DC: Department of the Army, 1961); US Army, *Special Forces Operational Techniques: FM 31-20,* (Washington, DC: Department of the Army, 1965); US Army, *Special Forces Operations: FM 31-21* (Washington, DC: Department of the Army, 1965); US Army, *Special Forces Operations: FM 31-21* (Washington, DC: Department of the Army, 1969).

9. See McClintock, *Instruments of Statecraft.*

10. As quoted in McClintock, *Instruments of Statecraft,* 219; emphasis added.

11. Ervand Abrahamian, *Iran Between Two Revolutions* (Princeton, NJ: Princeton University Press, 1982), 55–56.

12. Ervand Abrahamian, "The 1953 Coup in Iran," *Science & Society* (2001), 182–215; Ervand Abrahamian, *The Coup: 1953, the CIA, and the Roots of Modern US-Iranian Relations* (New York: The New Press, 2013). Darioush Bayandor, *Iran and the CIA: The Fall of Mosaddeq Revisited* (London: Palgrave Macmillan, 2010); Mark J. Gasiorowski and Malcolm Byrne, *Mohammad Mosaddeq and the 1953 Coup in Iran* (Syracuse, NY: Syracuse University Press, 2004); Kinzer, *Overthrow*; Kinzer, S., *All the Shah's Men: An American Coup and the Roots of Middle East Terror* (Oxford: John Wiley & Sons, 2003).

13. CIA, *The Battle for Iran* (released in 1981), in Malcolm Byrne, ed., *CIA Confirms Role in 1953 Iran Coup.* August 19, 2013. http://nsarchive.gwu.edu/NSAEBB/NSAEBB435/ (accessed October 27, 2016); Wilber, Donald. *Overthrow of Premier Mossadeq of Iran November 1952–August 1953,* CIA Clandestine Service History. CIA Historical Paper no 208 (1954, published 1969), in Byrne, ibid.

14. Doyle, Kate and Carlos Osorio, *US Policy in Guatemala, 1966–1996.* National Security Archive Electronic Briefing Book No. 11: The National Security Archive, n.d.; Alejandro de Quesada, *The Bay of Pigs* (Oxford: Osprey, 2009); Howard Jones, *The Bay of Pigs* (Oxford: Oxford University Press, 2008); Peter Kornbluh, *Bay of Pigs Declassified: The Secret CIA Report on the Invasion of Bay of Pigs* (New York: New Press, 1998); Jim Rasenberger, *The Brilliant Disaster: JFK, Castro, and America's Doomed Invasion of Cuba's Bay of Pigs* (New York: Simon and Schuster, 2012).

15. Rasenberger, *The Brilliant Disaster*; Kornbluh, *Bay of Pigs Declassified: The Secret CIA Report on the Invasion of Bay of Pigs*; Colonel J. Hawkins, *Record of Paramilitary Action Against the Castro Government of Cuba 17 March 1960–May 1961* (Washington, DC: CIA, 1961).

16. Stephen Schlesinger and Stephen Kinzer, *Bitter Fruit: The Untold Story of the American Coup in Guatemala* (Garden City, NY: Anchor Books, 1983) Kinzer, *Overthrow,* 134.

17. Manucher Farmanfarmaian and Roxane Farmanfarmaian, *Blood and Oil: Inside the Shah's Iran.* (New York: Modern Library, 1999).

18. Church Committee Reports, *United States Senate Select Committee to Study Governmental Operations with Respect to Intelligence Activities* (Washington, DC: United States Senate, 1975), III.3; Simon Collier and William Sater, *A History of Chile, 1808–2002*. (Cambridge: Cambridge University Press, 2004), 334–349.

19. Church Committee Reports, *United States Senate Select Committee to Study Governmental Operations with Respect to Intelligence Activities*, II.15; Collier and Sater, *A History of Chile*, 334–349.

20. Gregory Palast, "A Marxist Threat to Cola Sales? Pepsi Demands a US Coup. Goodbye Allende. Hello Pinochet" *The Observer*, November 8, 1998. www.theguardian.com/business/1998/nov/08/observerbusiness.theobserver

21. Church Committee Reports *United States Senate Select Committee to Study Governmental Operations with Respect to Intelligence Activities*.

22. CIA, "Notes on Meeting with the President on Chile," September 15, 1970; Peter Kornbluh, ed., *Chile and the United States: Declassified Documents Relating to the Military Coup, September 11, 1973*. National Security Archive Electronic Briefing Book No. 8. http://nsarchive2.gwu.edu//NSAEBB/NSAEBB8/nsaebb8i.htm

23. CIA, "Operating Guidance Cable on Coup Plotting." CIA memorandum/cable, 1970; Kornbluh, ed., *Chile and the United States*.

24. US Office of the President, "Memorandum of a Conference With the President, White House, Washington, DC, March 29, 1957, 11:15 a.m.," US Office of the Historian. Foreign Relations of the United States 1955–1957, National Security Policy, Volume XIX. https://history.state.gov/historical documents/frus1955-57v19/d113

25. Michael A. Innes, ed., *Making Sense of Proxy Wars: States, Surrogates, and the use of Force* (Dulles, VA: Potomac Books, 2012); Michael T. Klare, "Subterranean Alliances: America's Global Proxy Network," *Journal of International Affairs*, 43, no. 1 (1989), 97–118; Kinzer, *Overthrow*, 5.

26. Prados, *Safe for Democracy*, 148.

27. US National Security Council 1948, *National Security Council Directive NSC 10/2 June 18, 1948* (Washington, DC: US National Security Council), June 18, 1948.

28. McClintock, *Instruments of Statecraft*, 28.

29. US Army, *Guerrilla Warfare and Special Forces Operations FM 31-21* (Washington, DC: Headquarters, Department of the Army, 1958), 20.

30. Haynes Johnson, *The Bay of Pigs* (London: Hutchinson and Co., 1965), 9.

31. Arthur Schlesinger, "Memorandum From the President's Special Assistant (Schlesinger) to President Kennedy," Washington, DC, April 10, 1961, in US Office of the Historian, *Foreign Relations of the United States, 1961–1963, Volume X, Cuba, January 1961–September 1962*. https://history.state.gov/historicaldocuments/frus1961-63v10/d86, 8.

32. George Kennan, as quoted in McClintock, *Instruments of Statecraft*, 28.

33. US Office of the President, *United States Overseas Internal Defense Policy* (Washington, DC: National Security Action Memorandum 182, 1962), 10.

34. CIA, *Power Moves*, 18.

35. Schlesinger, Arthur, "Memorandum From the President's Special Assistant [Schlesinger] to President Kennedy," Washington, DC, April 10, 1961, in Office of the Historian, *Foreign Relations of the United States, 1961–1963, Volume X, Cuba, January 1961–September 1962*. https://history.state.gov/ historicaldocuments/frus1961-63v10/d86

36. Colonel J. Hawkins, *Record of Paramilitary Action Against the Castro Government of Cuba 17 March 1960–May 1961* (Washington, DC: Clandestine Services History, CS Historical Paper no. 105, 1961). https://fas. org/irp/cia/product/cuba.pdf (accessed November 18, 2017).

37. Schlesinger, *Memorandum for the President*, 2.

38. See, for example, John Carter, *Covert Action as a Tool of Presidential Foreign Policy: From the Bay of Pigs to Iran-Contra* (Lewiston, NY: Edwin Mellen Press, 2006), 223.

39. Prados, *Presidents' Secret Wars*, 188, 194; Klaas Voß, "Plausibly Deniable: Mercenaries in US Covert Interventions During the Cold War, 1964–1987," *Cold War History*, 16, no. 1. (2016), 40.

40. Kenneth J. Conboy and James Morrison, *Feet to the Fire: CIA Covert Operations in Indonesia, 1957–1958* (Annapolis, MD: Naval Institute Press, 1999); see also Prados, *Safe for Democracy*, 178–180.

41. CIA, *Power Moves*, 23.

42. US Army, *Special Forces Operations: FM 31-21*.

43. CIA. *Psychological Operations in Guerrilla Warfare* (Langley, VA: CIA, 1983). www.cia.gov/library/readingroom/docs/CIA-RDP86M00886R0013000 10029-9.pdf; CIA, *The Freedom Fighter's Manual* (Langley, VA: CIA, 1983). http://destructables.org/sites/default/files/destructable/step/downloads/ FreedomFightersManual.pdf

44. US Army, *Guerrilla Warfare and Special Forces Operations FM 31-21* (Washington, DC: Department of the Army, 1961).

45. CIA, *Power Moves*, 38.

46. US Department of State, "Measures which the United States Government Might Take in Support of a Successor Government to Mossadeq," Top Secret Memorandum, March 1953, in Mark J. Gasiorowski and Malcolm Byrne, eds., *Mohammed Mossadeq and the 1953 Coup in Iran*, The National Security Archive, June 22, 2004. https://nsarchive2.gwu.edu/NSAEBB/ NSAEBB126/index.htm

47. Ali Rahnema, *Behind the 1953 Coup in Iran: Thugs, Turncoats, Soldiers, and Spooks* (Cambridge: Cambridge University Press, 2014), 4–7; Abrahamian, "The 1953 Coup in Iran," 212–213; also see Abrahamian, *The Coup: 1953, the CIA, and the Roots of Modern US-Iranian Relations*.

48. Church Committee Reports ,*United States Senate Select Committee to Study Governmental Operations with Respect to Intelligence Activities.*

49. Church Committee Reports *United States Senate Select Committee to Study Governmental Operations with Respect to Intelligence Activities*, 31.

50. Wilber, Donald, *Overthrow of Premier Mossadeq of Iran November 1952– August 1953*. CIA Clandestine Service History. CIA Historical Paper no. 208, 1954 (published 1969). Available online at National Security Archive: Malcolm Byrne, ed., "CIA Confirms Role in 1953 Iran Coup," August 19,

2013. National Security Archive Electronic Briefing Book no. 435. http://nsarchive.gwu.edu/NSAEBB/NSAEBB435/ (accessed October 27, 2016); Mark J. Gasiorowski, "The 1953 Coup d'Etat in Iran," *International Journal of Middle East Studies* 19, no. 3 (1987), 261–286; Abrahamian, "The 1953 Coup in Iran," 182–215; Bayandor, *Iran and the CIA: The Fall of Mosaddeq Revisited*.

51. See, for example, Byrne, ed., "CIA Confirms Role in 1953 Coup."

52. Rahnema, *Behind the 1953 Coup in Iran*, 44.

53. US Army, *Guerrilla Warfare and Special Forces Operations FM 31-21* (Washington, DC: Department of the Army, 1958), 18.

54. US Army, *Special Forces Operations: FM 31-21* (Washington, DC: Department of the Army, 1969), 3–9.

55. CIA, *Power Moves*, 21; US Army, *Special Forces Operational Techniques: FM 31-20* (Washington, DC: Department of the Army, 1965); US Army, *Unconventional Warfare Devices and Techniques: Incendiaries TM 31-201-1* (Washington, DC: Department of the Army, 1966).

56. CIA, *Power Moves*, 31.

57. Stathis Kalyvas, *The Logic of Violence in Civil War* (Cambridge: Cambridge University Press, 2006, reprinted 2009).

58. CIA, *Power Moves*, 16.

59. As quoted in William M. Leary, "CIA Air Operations in Laos 1955–1974: Supporting the Secret War," CIA Webpage, 2007 (last updated 2008). www.cia.gov/library/center-for-the-study-of-intelligence/csi-publications/csi-studies/studies/winter99-00/art7.html (accessed September 4, 2016); Christopher Robbins, *The Ravens: Pilots of the Secret War of Laos* (New York: Crown Publishers, 1987).

60. Ganser, *NATO's Secret Armies: Operation Gladio and Terrorism in Western Europe*; Paul L. Williams, *Operation Gladio: The Unholy Alliance between the Vatican, the CIA and the Mafia* (New York: Prometheus Books, 2015); Leopoldo Nuti, "The Italian 'Stay-Behind' Network—The Origins of Operation 'Gladio,'" *Journal of Strategic Studies*, 30, no. 6 (2007), 955–980.

61. Ganser, *NATO's Secret Armies*, 2.

62. Prados, *Presidents' Secret Wars*, 30–60.

63. Peter Schraeder, "Paramilitary Intervention," in idem, ed., *Intervention into the 1990s* (London: Lynne Rienner Publishers, 1992), 131–152.

64. Voß, "Plausibly Deniable"; Guy Arnold, *Mercenaries: The Scourge of the Third World* (New York: St. Martin's Press, 1999); Jay Mallin and Robert K. Brown, *Merc: American Soldiers of Fortune* (New York: Macmillan Pub. Co., 1980); Wilfred Burchett and Derek Roebuck, *The Whores of War: Mercenaries Today* (New York: Penguin Books, 1977).

65. Air America (henceforth AA), the principal CIA propriety airline, was often the umbrella name given to a conglomeration of CIA propriety and/or sponsored airlines that the CIA had at its disposal. Other airlines used by the CIA included Civil Air Transport (CAT), Intermountain Aviation, Southern Air Transport, and Bird Airlines; even Air Ethiopia, Air Jordan, and Iran Air were some of the airlines the CIA subsidized around the world. See Robbins, *Air America*, 47.

66. Stephen Grey, *Ghost Plane* (London: Hurst and Company, 2006), 95–97; Scott Shane, Stephen Grey and Margot Williams, "C.I.A. Expanding Terror Battle Under Guise of Charter Flights," *New York Times*, May 31, 2005.

67. William M. Leary and William Stueck, "The Chennault Plan to Save China: US Containment in Asia and the Origins of the CIA's Aerial Empire, 1949–1950," *Diplomatic History* 8, no. 4 (1984), 349–364: Leary, *CIA Air Operations in Laos 1955–1974*; William Leary, *Perilous Missions: Civil Air Transport and CIA Covert Operations in Asia* (Tuscaloosa, AL: University of Alabama Press, 2006, second ed.).

68. CIA, "Air America: Upholding the Airmen's Bond," CIA Website, Homepage of CIA FOIA, 2010. www.foia.cia.gov/collection/air-america-upholding-airmens-bond (accessed December 9, 2015); Prados, *Presidents' Secret Wars*, 62.

69. Robbins, *Air America*, 47.

70. Evan Thomas, "In Arizona: A Spymaster Remembered," *Time Magazine* (April 7, 1986).

71. As quoted in Robbins, *Air America*, 173.

72. Fred C. Moor, *Then Came the CIA: The Early Years of Southern Air Transport* (Create Space Independent Publishing, 2011).

73. Robbins, *Air America*, 47–54.

74. "Sheep-dipping" refers to a process of "civilianizing" a member of the armed forces, often involving the creation of false retirement documents, so that they can operate in a "private" capacity while denying they are members of the US military; see Robbins, *Air America*, 1; Holly Sklar, *Washington's War on Nicaragua* (Cambridge, MA: South End Press, 1988), 258. Such a process was used for many of the Air America and affiliated airline companies.

75. Robbins, *Air America*, 12.

76. P. W. Singer, *Corporate Warriors* (Ithaca, NY: Cornell University Press, 2003), 120.

77. CIA, *Power Moves Involved in the Overthrow of an Unfriendly Government*, 33, 47.

78. Robbins, *Air America*; Joe F. Leeker, *The History of Air America* (Dallas, TX: University of Texas Press, 2015, second ed.); Leary, *Perilous Missions*; Prados, *Presidents' Secret Wars*.

79. Church Committee Reports, *United States Senate Select Committee to Study Governmental Operations with Respect to Intelligence Activities*, Book 1, 206.

80. As quoted in Kinzer, *Overthrow*, 132.

81. Nick Cullather, *Secret History: The CIA's Classified Account of its Operations in Guatemala, 1952–1954* (Stanford, CA: Stanford University Press, 1999).

82. Schlesinger and Kinzer, *Bitter Fruit*, 8–9.

83. Richard H. Immerman, *The CIA in Guatemala: The Foreign Policy of Intervention* (Dallas, TX: University of Texas Press, 2010), 162.

84. Kate Doyle and Peter Kornbluh, *CIA and Assassinations: The Guatemala 1954 Documents*, National Security Archive Electronic Briefing Book, No.

4, The National Security Archive, n.d. http://nsarchive.gwu.edu/NSAEBB/NSAEBB4/

85. Haines, Gerald, *CIA and Guatemala Assassination Proposals 1952–1954. CIA History Staff Analysis,* CIA report (1995).

86. Thomas Bodenheimer and Robert Gould, *Rollback: Right Wing Power in US Foreign Policy* (Boston, MA: South End Press, 1989), 26; Leeker, *The History of Air America.*

87. Cullather, *Secret History,* 71; Leeker, *The History of Air America*; Bodenheimer and Gould, *Rollback: Right Wing Power in US Foreign Policy,* 73.

88. Immerman, *The CIA in Guatemala,* 3–5; see also, for example, Michael McClintock, *The American Connection, Volume 2: State Terror and Popular Resistance in Guatemala,* (London: Zed Books, 1985). Robert Parry, "History of Guatemala's Death Squads," *Consortium News,* January 11, 2005.

89. Audrey Kahin and George M. Kahin, *Subversion as Foreign Policy: The Secret Eisenhower and Dulles Debacle in Indonesia* (New York: The New Press, 1995); James Callanan, *Covert Action in the Cold War: U.S. Policy, Intelligence and CIA Operations* (New York: I.B. Taurus, 2010), 133–136; Prados, *Safe For Democracy,* 170–180; Matthew Jones, "'Maximum disavowable aid': Britain, the United States and the Indonesian Rebellion, 1957–58," *English Historical Review,* 114, no. 459 (1999): 1179–1216.

90. Jones, *The Bay of Pigs,* 9.

91. Jones, *The Bay of Pigs,* 10; Quesada, *The Bay of Pigs*; William Stodden, and Ari Weiss, "Interests and Foreign Policy: The Cuban Revolution and US Response, 1959–1961," *Foreign Policy Analysis,* 13, no. 1 (2016), 74–92.

92. Prados, *Presidents' Secret Wars,* 180.

93. Hawkins, *Record of Paramilitary Action,* 12; Leeker, *The History of Air America,* "Air America at Bay of Pigs,", 8–9).

94. 5412 Committee, *A Program of Covert Action Against the Castro Regime* (Washington, DC: Department of State, 1960).

95. Prados, *Presidents' Secret Wars,* 184; Jones, *The Bay of Pigs.*

96. Lauren Harper and Thomas Blanton, *CIA Releases Controversial Bay of Pigs History,* National Security Archive Electronic Briefing Book no. 564: The National Security Archive, October 31, 2016. https://nsarchive2.gwu.edu/NSAEBB/NSAEBB564-CIA-Releases-Controversial-Bay-of-Pigs-History/

97. Hawkins, *Record of Paramilitary Action,* 7, 13, 43; Jones, *The Bay of Pigs,* 115.

98. Leeker, *The History of Air America,* "Air America at Bay of Pigs", 3; Hawkins, *Record of Paramilitary Action,* 12; Moor, *Then Came the CIA.*

99. Quesada, *The Bay of Pigs*; Prados, *Presidents' Secret Wars,* 184, 195.

100. Jones, *The Bay of Pigs* 101, 131.

101. Prados, *Presidents' Secret Wars,* 231.

102. Thomas, *In Arizona: A Spymaster Remembered.*

103. Ibid.

104. Peter Kornbluh, "Luis Posada Carriles: The Declassified Record," National Security Archive, Electronic Briefing Book No. 153, The National

Security Archive, May 10, 2005. https://nsarchive2.gwu.edu/NSAEBB/
NSAEBB153/index.htm; Peter Kornbluh, "The Posada File: Part II,"
National Security Archive Electronic Briefing Book No. 157, The National
Security Archive, June 9, 2005. https://nsarchive2.gwu.edu/NSAEBB/
NSAEBB157/index.htm

105. Tim Weiner, "Cuban Exile Could Test US Definition of Terrorist," *New York Times*, May 9, 2005.

106. Peter Kornbluh and Erin Maskell, "The CIA File on Luis Posada Carriles," National Security Archive Electronic Briefing Book No. 334: The National Security Archive, January 11, 2011. https://nsarchive2.gwu.edu/NSAEBB/NSAEBB334/index.htm

107. Kornbluh, "Luis Posada Carriles: The Declassified Record," Documents 14 and 15.

108. Kornbluh, "The Posada File: Part II," Document 12; Kornbluh and Maskell, "The CIA File on Luis Posada Carriles."

109. US Department of Justice, *Omega 7* (Washington, DC: US Department of Justice and FBI [1993]).

110. Leary, *CIA Air Operations in Laos 1955–1947*; Timothey Castle, *At War in the Shadow of Vietnam: United States Military Aid to the Royal Lao Government, 1955–75* (New York: Columbia University Press, 1995).

111. Prados, *Presidents' Secret Wars*, 268.

112. US Department of Defense, "United States-Vietnam Relations 1945–1967," Study Prepared by the Department of Defense, Book 10 of 12, 1971. www.dtic.mil/get-tr-doc/pdf?AD=ADA542353; Thomas L. Ahern, Jr., *Undercover Armies: CIA and Surrogate Warfare in Laos: 1961–1973* (Washington, DC: Central Intelligence Agency, 2006); Kenneth Conboy, *Shadow War: The CIA's Secret War in Laos* (Boulder, CO: Paladin Press, 1995).

113. Prados, *Presidents' Secret Wars*, 261–269; Ahern, *Undercover Armies: CIA and Surrogate Warfare in Laos: 1961–1973*; Conboy, *Shadow War: The CIA's Secret War in Lao.*

114. Alan Axelrod, *Mercenaries: A Guide to Private Armies and Private Military Companies* (Thousand Oaks, CA: CQ Press, 2014).

115. Joshua Kurlantzick, *A Great Place to Have a War: America in Laos and the Birth of a Military CIA* (New York: Simon and Shuster, 2017), 128, 213, 261.

116. Leeker, *The History of Air America*; Robbins, *Air America*, 113.

117. Ahern, *Undercover Armies*, 191; Laurie Clayton, "CIA and the Wars in Southeast Asia 1947–1975," *Studies in Intelligence* (Washington, DC: CIA, August 2016. file:///C:/Users/3048017/Downloads/vietnam-anthology-print-version.pdf

118. James Dunigan and Albert Nofi, *Dirty Little Secrets of the Vietnam War* (New York: St. Martin's Press, 1999), 185.

119. Prados, *Presidents' Secret Wars*, 341–348, 357–360.

3. Counterinsurgent Statecraft: Militias, Mercenaries, and Contractors

1. US Office of the President, *United States Overseas Internal Defense Policy*, National Security Action Memorandum 182, Washington, DC, 1962.
2. See notes 69 and 70 below and in-text information. See Howard Jones, *A New Kind of War: America's Global Strategy and the Truman Doctrine in Greece* (Oxford: Oxford University Press, 1997), 137. Also see an analysis of paramilitary units in Greece: William Needham, "Paramilitary Forces in Greece, 1946–1949" (Carlisle Barracks, PA: Army War College, February 26, 1971). www.dtic.mil/dtic/tr/fulltext/u2/773486.pdf. Also see Andrew Birtle, *US Army Counterinsurgency and Contingency Operations Doctrine 1942–1976* (Washington, DC: Center of Military History United States Army, 2006); Chester J. Pach, *Arming the Free World: The Origins of the United States Military Assistance Program, 1945–1950* (Chapel Hill, NC: University of North Carolina Press Books, 1991).
3. US planners supported a network of "stay behind" paramilitary forces in case of a Soviet invasion or a communist takeover of Italy. When this failed to materialize, these forces were used as pro-government militias or a right-wing paramilitary force to intimidate the left-wing political parties and movements (including left-wing insurgent factions) in what is often referred to as the "years of lead," a period of intense political tension over the direction of Italian politics during the 1960s through to the 1980s. See, for example, Leopoldo Nuti, "The Italian 'Stay-Behind' Network—The Origins of Operation 'Gladio'," *Journal of Strategic Studies*, 30, no. 6 (2007), 955–980; Paul L. Williams, *Operation Gladio: The Unholy Alliance Between the Vatican, the CIA, and the Mafia* (New York: Prometheus Books, 2015). For original documentation in Italian, see Senato Della Repubblica, "Il Terrorismo, Le Stragi ed il Contesto Storico-Publico," Commissione Parlamentare D'inchiesta Sul Terrorismo in Italia E Sulle Cause Della Mancata Individualizone Dei Responsibili Delle Stragi, May 17, 1988. https://web.archive.org/web/20060819211212/http://www.isn.ethz.ch/php/documents/collection_gladio/report_ital_senate.pdf
4. US advisors in the Philippines mobilized and directed various paramilitary militias and recommended to the Filipino Army to do the same to fight against the Huk rebellion. See Napoleon Valeriano and Charles Bohannan, *Counter-Guerrilla Operations: The Philippine Experience* (Westport, CT: Praeger Security International, 2006 [1992]); US Army, "Counter-Guerrilla Operations in the Philippines, 1946–1953," Transcripts of a Seminar on the Huk Champaign (Fort Bragg, NC, June 15, 1961, 61. https://murdercube.com/files/Combined%20Arms/counterguerrillaops.pdf. Much later, US counterinsurgency experts and advisors helped the Filipinos develop various militia forces; see David Kowalewski, "Counterinsurgent Paramilitarism: A Philippine Case Study," *Journal of Peace Research*, 29, no. 1 (February, 1992), 71–84; Walden Bello, "Counterinsurgency's Proving Ground: Low-Intensity Warfare in the Philippines," in Michael T. Klare and Peter Kornbluh, eds., *Low Intensity Warfare: Counterinsurgency,*

Proinsurgency and Anti-Terrorism in the Eighties (New York: Pantheon Books, 1988), 158–182.

5. Daniel Fineman, *A Special Relationship: The United States and Military Government in Thailand 1947–1958* (Honolulu, HI: University of Hawaii Press, 1997), 182–184; International Crisis Group, *Southern Thailand: The Problem with Paramilitaries*, Asia Report 140, 2007; Arne Kislenko, "A Not so Silent Partner: Thailand's Role in Covert Operations, Counter-Insurgency, and the Wars in Indochina," *Journal of Conflict Studies*, 24, no. 1 (2004).

6. See Birtle, *US Army Counterinsurgency*, 93; Church Committee Reports, *United States Senate Select Committee to Study Governmental Operations with Respect to Intelligence Activities* (Washington, DC: United State Senate, 1975), "Foreign and Military Intelligence," 23–28.

7. See, for example, Michael McClintock, *The American Connection, Volume 2: State Terror and Popular Resistance in Guatemala* (London: Zed Books, 1985); Susanne Jonas, *The Battle for Guatemala: Rebels, Death Squads, and US Power* (Boulder, CO: Westview Press, 1991); Robert Parry, "History of Guatemala's Death Squads," *Consortium News*, January 11, 2005.

8. The US provided military assistance and advice to Indonesia's new government and armed forces after a coup overthrew the left-wing Sukarno regime. For the US's role in the coup and US fears of the left-wing Sukarno government and its ties to the communists, see Peter Dale Scott, "The United States and the Overthrow of Sukarno, 1965–1967," *Pacific Affairs*, 58, no. 2 (1985), 239–264. The Indonesian Army and paramilitary forces began to eliminate the civilian left, brutally murdering hundreds of thousands of civilians. Declassified documents demonstrate the US was not only aware but also complicit in this mass murder. See Brad Simpson, *U.S. Embassy Tracked Indonesia Mass Murder 1965*. National Security Archive Electronic Briefing Book No. 607, National Security Archive, October 17, 2017. Documents revealed how the US offered covert support to the Indonesian military during the massacres. They also revealed the Indonesian Army and paramilitary forces' targeting of unions and labor groups of employees of major oil companies in a bid to weaken "communist infiltration" and collective action. Also see Brad Simpson, *Economists with Guns: Authoritarian Development and US-Indonesian Relations, 1960–1968* (Stanford, CA: Stanford University Press, 2008). For putting this in the context of world order and capital accumulation, see Hilmar Farid, "Indonesia's Original Sin: Mass Killings and Capitalist Expansion, 1965–66," *Inter-Asia Cultural Studies*, 6, no. 1 (2005), 3–16.

9. See the brief overview in this chapter.

10. See the brief overview in this chapter and Chapter 5.

11. After the US-supported coup in Chile in 1973, which is often referred to as the "other 9/11," the new US-backed Pinochet government embarked on an economic liberalization program. This included training from the "Chicago boys" (economists led by Milton Friedman at the University of Chicago) and a US-supported counterinsurgent security apparatus to contain dissent and subversion of the newly imposed order. Chilean

officials encouraged the formation of semi-official and plausibly deniable militias and death squads as part of this counterinsurgent security apparatus. For more on the semi-official militias, see the "Caravan of Death" military unit in the references below, and for informal or deniable death squads, please see what some of the reference sources below refer to as "clandestine forces" and "September 11 commandos": Mark Ensalaco, *Chile Under Pinochet: Recovering the Truth* (Philadelphia, PA: University of Pennsylvania Press, 2000); John Dinges, *The Condor Years: How Pinochet and His Allies Brought Terrorism to Three Continents* (New York: The New Press, 2005); Peter Kornbluh, *The Pinochet File: A Declassified Dossier on Atrocity and Accountability* (New York: The New Press, 2013). In addition, for more information on how certain groups, such as the Argentine Anti-communist Alliance (AAA) which formed part of the transnational Latin American counterinsurgent network called "Operation Condor," see Thomas Wright, *State Terrorism in Latin America: Chile, Argentina, and International Human Rights* (Plymouth: Rowman and Littlefield Publishing Group, 2007). For Operation Condor, and Chile's participation, see Patrice McSherry, *Predatory States: Operation Condor and Covert War in Latin America* (Lanham, MD: Rowman & Littlefield, 2005). For more on the informal militias, see the evidence for coding in the entry on Chile in Sabine Carey, Neil Mitchell and Will Lowe, "States, the Security Sector and the Monopoly of Violence: A New Database on Pro-Government Militias," *Journal of Peace Research*, 50, no. 2 (March, 2013), 249–258, and the Pro-Government Militias Database (PGMD) http://www.sabinecarey.com/militias/

12. McSherry, *Predatory States*; Patrice McSherry, "Tracking the Origins of a State Terror Network: Operation Condor," *Latin American Perspectives* (2002), 38–60.

13. For a longer list of similar counterinsurgency support with military-paramilitary connections, see William Blum, *Rogue State: A Guide to the World's Only Superpower* (London: Zed Books, 2003, second ed.), 122–167.

14. See, for example, US Army, *Operations Against Irregular Forces: FM 31-15* (Washington, DC: Headquarters, Department of the Army, 1961); US Army, *Special Forces Operational Techniques: FM 31-20* (Washington, DC: Headquarters, Department of the Army, 1965); US Army, *Counterguerrilla Operations: FM 90-8* (Washington, DC: Headquarters, Department of the Army, 1986); US Army, *Doctrine for Special Forces Operations FM 31-20* (Washington, DC: Department of the Army, 1990).

15. US Army, *Counterguerrilla Operations: FM 31-16*, 40; US Army, *US Army Counterinsurgency Forces: FM 31-22* (Washington, DC: Department of the Army, 1963), 19 and 82; US Army, *Stability Operations: FM 31-23* (Washington, DC: Department of the Army, 1972), Section 8.

16. US Army, *Special Forces Operational Techniques: FM 31-20*, 70.

17. US Army, *Counterguerrilla Operations: FM 31-16* (Washington, DC: Department of the Army, 1967), 4, 40.

18. Michael McClintock, *Instruments of Statecraft: U.S. Guerrilla Warfare,*

Counterinsurgency, and Counterterrorism, 1940–1990 (New York: Pantheon Books, 1992), 44.

19. US Army, *Stability Operations: FM 31-23* (Washington, DC: Department of the Army, 1972), 1-1.

20. US Army, *Operations Against Irregular Forces: FM 31-15* (Washington, DC: Headquarters, Department of the Army, 1961), 4–6.

21. Lisa Haugaard, "Declassified Army and CIA Manuals used in Latin America: An Analysis of their Content," *Latin America Working Group*, 1997; Doug Stokes and Sam Raphael, *Global Energy Security and American Hegemony* (Baltimore, MD: John Hopkins University Press, 2010), 65; Michael McClintock, "American Doctrine and Counterinsurgent State Terror," in Alexander George, ed., *Western State Terrorism* (Oxford: Polity Press, 1991); McClintock, *Instruments of Statecraft*.

22. US Army, *Operations in a Low Intensity Conflict: FM 7-98* (Washington, DC: Department of the Army, 1992), Section 1.

23. Haugaard, *Declassified Army and CIA Manuals.*

24. US Army, *Operations Against Irregular Forces*, 65.

25. *Revolutionary War, Guerrillas and Communist Ideology* (1989) (p. 51—as quoted in Haugaard, *Declassified Army and CIA Manuals.*

26. US Army, *Stability Operations: FM 31-23*, 3-4.

27. US Army, *U.S. Army Counterinsurgency Forces: FM 31-22*, 4.

28. US Army, *Counterguerrilla Operations: FM 90-8* Appendix H.

29. US Army, *Operations in a Low Intensity Conflict: FM 7-98* (Washington, DC: Department of the Army, 1992).

30. McClintock, *American Doctrine and Counterinsurgent State Terror*, 125; Gonzalo Sánchez and Donny Meertens, *Bandits, Peasants, and Politics*, trans. Alan Hynds (Austin, TX: University of Texas Press, 2001).

31. Haugaard, *Declassified Army and CIA Manuals*; US Army, *Operations in a Low Intensity Conflict: FM 7-98*, 2-27.

32. US Army, *Operations Against Irregular Forces: FM 31-15*, 14.

33. Comando del Ejercito, *Reglamento de Combate de Contraguerrilleras EJC 3-10* (Bogota: Ejercito de Colombia, 1987), 197.

34. US Army, *US Army Counterinsurgency Forces: FM 31-22*, 106

35. US Army, *Counterguerrilla Operations: FM 31-16*, 12; US Army, *Operations Against Irregular Forces: FM 31-15*, 12.

36. McClintock, *Instruments of Statecraft*, 28.

37. US Army, *Handbook for Aggressor Insurgent War, FM 30-104* (Washington, DC, September, 1967), 3.

38. Ninth Psychological Warfare Company. "Program of Instruction," a Ninth Psychological Warfare Company Document (Fort Gulick, Panama Canal Zone: circa 1966). The 9th Psychological Warfare Company, later renamed the 9th Psychological Warfare Battalion in 1967, operated out of Fort Gulick in the Panama Canal Zone as part of the US Special Warfare Center located there between 1963 and 1974. For more details on the courses offered there, including use of the term "propaganda of the deed," see Willard Barber and C. Neal Ronning, *Internal Security and Military Power: Counterinsurgency and Civic Action in Latin America* (Columbus,

OH: Ohio State University Press, 1966). "Propaganda of the deed" was a term US military planners often used in the early 1960s to refer to actions (not necessarily violent) that would have a psychological effect. See Carnes Lord and Frank Barnett, *Political Warfare and Psychological Operations* (Washington, DC: National Strategy Information Center, 1989).

39. CIA, *Psychological Operations in Guerrilla Warfare* (Langley, VA: CIA, 1983). www.cia.gov/library/readingroom/docs/CIA-RDP86M00886R0013000 10029-9.pdf

40. US Army, *Operations Against Irregular Forces: FM 31-15*, 47.

41. US Army, *US Army Handbook of Counterinsurgency Guidelines for Area Commanders: An Analysis of Criteria, No. 550-100* (Washington, DC: Department of the Army, 1966), 225.

42. US Marine Corps, *Counterguerrilla Operations: MCRP 3-33A* (Washington, DC: US Government Printing Office, 1986), 3-5.

43. US Army, *Counterguerrilla Operations: FM 31-16*, 133.

44. McClintock, *Instruments of Statecraft*.

45. See, for example, US Army, *U.S. Army Counterinsurgency Forces: FM 31-22*.

46. David Galula, *Counterinsurgency Warfare: Theory and Practice* (New York: Praeger, 1964), 124.

47. US Army, *Counterguerrilla Operations: FM 31-16* (Washington, DC: Department of the Army, 1967), 40.

48. US Army, *Stability Operations: FM 31-23* (Washington, DC: Department of the Army, 1972), 8-8.

49. US Army, *Operations Against Irregular Forces: FM 31-15*, 36.

50. US Navy/Marine Corps, *Operations Against Guerrilla Forces FM 21* (Washington, DC: Department of Marine Corps, 1962, 1965), 74.

51. US Army, *Counterguerrilla Operations: FM 31-16*.

52. McClintock, *Instruments of Statecraft*, 243.

53. US Army, *Counterguerrilla Operations: FM 31-16*, 13–16.

54. US Army, *U.S. Army Counterinsurgency Forces: FM 31-22*, 84.

55. Peter Paret and John Shy, "Guerrilla Warfare and U.S. Military Policy: A Study," in US Navy, ed., *The Guerrilla and How to Fight Him FMFRP 12-25* (Washington, DC: US Marine Corps, 1962, reprinted 1990), 52.

56. Roger Trinquier, *Modern Warfare: A French View of Counterinsurgency* (Westport, CT: Praeger, 1964), 105.

57. Govinda Clayton and Andrew Thomson, "Civilianizing Civil Conflict: Civilian Defense Militias and the Logic of Violence in Intrastate Conflict," *International Studies Quarterly*, 60, no. 3 (2016), 499–510; Goran Peic, "Civilian Defense Forces, State Capacity, and Government Victory in Counterinsurgency Wars," *Studies in Conflict and Terrorism*, 37, no. 2 (2014), 162–184; Metin Gurcan, "Arming Civilians as a Counterterror Strategy: The Case of the Village Guard System in Turkey," *Dynamics of Asymmetric Conflict*, 8, no. 1 (2015), 1–22; Jason Lyall, "Are Coethnics More Effective Counterinsurgents? Evidence from the Second Chechen War," *American Political Science Review*, 104, no. 1 (2010).

58. US Army, *Special Forces Operations: FM 31-21* (Washington, DC: Department of the Army, 1969), 10-2.

59. National Security Council, National Security Action Memorandum (NSAM 2), "Development of Counter-Guerrilla Forces" to the Secretary of Defense Robert S. McNamara from McGeorge Bundy, Special Assistant to the President for National Security Affairs, February 3, 1961, Kennedy Library, National Security Council. www.jfklibrary.org/Asset-Viewer/ Archives/JFKNSF-328-003.aspx

60. National Security Council, National Security Action Memorandum (NSAM 56), "National Security Action Memorandum Number 56: Evaluation of Paramilitary Requirements," June 28, 1961, Kennedy Library, National Security Council. www.jfklibrary.org/Asset-Viewer/ ex2GImrvWU-Z9Zqyga7ryQ.aspx; National Security Council, National Security Action Memorandum (NSAM 162), "Development of U.S. and Indigenous Police, Paramilitary and Military Resources," June 19, 1962, Kennedy Library, National Security Council. www.jfklibrary.org/Asset-Viewer/e2A7V5fk8oajVcTlZCkxxw.aspx

61. As quoted in McClintock, *Instruments of Statecraft*, 163.

62. See, for example, Ruth Blakeley, "Still Training to Torture? US Training of Military Forces from Latin America," *Third World Quarterly*, 27, no. 8 (2006), 1439–1461; Jeremy Kuzmarov, "Modernizing Repression: Police Training, Political Violence, and Nation-Building in the 'American Century,'" *Diplomatic History*, 33, no. 2 (2009), 191–221; Jesse Savage and Jonathan Caverley, "When Human Capital Threatens the Capitol: Foreign Aid in the Form of Military Training and Coups," *Journal of Peace Research*, 54, no. 4 (2017), 542–557.

63. *Operaciones de Contraguerrilla FM 31-16* (1968), quoted in McClintock, *Instruments of Statecraft*, 244.

64. Ariel Ahram, *Proxy Warriors* (Stanford, CA: Stanford University Press, 2011); Joel Migdal, *Strong Societies and Weak States: State-Society Relations and State Capabilities in the Third World* (Princeton, NJ: Princeton University Press, 1988); Phillip Khoury and Joseph Kostiner, eds., *Tribes and State Formation in the Middle East* (Santa Barbara, CA: University of California Press, 1990); Hillel Frisch "Explaining Third World Security Structures," *Journal of Strategic Studies*, 25, no./ 3 (2002), 161–190.

65. Noam Chomsky and Edward Herman, *The Washington Connection and Third World Fascism* (Boston, MA: South End Press, 1979); Edward Herman, *The Real Terror Network* (Montreal: Black Rose Books, 1982); Michael T. Klare and Cynthia Arnson, *Supplying Repression: US Support for Authoritarian Regimes Abroad* (Washington, DC: Institute for Policy Studies, 1981); Lesley Gill, *The School of the Americas: Military Training and Political Violence in the Americas* (Durham, NC: Duke University Press, 2004); Grandin, *Empire's Workshop*.

66. Birtle, *US Army Counterinsurgency*.

67. Emile Simpson, *War from the Ground Up: Twenty-First Century Combat as Politics* (Oxford: Oxford University Press, 2012), 81–82; William Blum,

Killing Hope: US Military and CIA Interventions Since WW2 (London: Zed Books, 2002), 35–37.

68. See Howard Jones, *A New Kind of War: America's Global Strategy and the Truman Doctrine in Greece* (Oxford: Oxford University Press, 1997), 137. Also see an analysis of paramilitary units in Greece: William Needham, "Paramilitary Forces in Greece, 1946–1949 (Carlisle Barracks, PA: Army War College, February 26, 1971). www.dtic.mil/dtic/tr/fulltext/u2/773486. pdf. Also see Andrew Birtle, *US Army Counterinsurgency*; Pach, *Arming the Free World*.

69. Benedict Kerkvliet, *The Huk Rebellion: A Study of Peasant Revolt in the Philippines* (New York: Rowman & Littlefield Publishers, Inc., 2002).; See also Stanley Karnow, *In Our Image: America's Empire in the Philippines* (New York: Ballantine Books, 1989).

70. Napoleon Valeriano and Charles Bohannan, *Counter-Guerrilla Operations: The Philippine Experience* (Westport, CT: Praeger Security International, 2006 [1992]).

71. Stephen Shalom, "Counter Insurgency in the Philippines," in Daniel Schirmer and Stephen Shalom, eds., *The Philippines Reader: A History of Colonialism, Neocolonialism, Dictatorship and Resistance* (Boston, MA: South End Press, 1987); Valeriano and Bohannan, *Counter-Guerrilla Operations*; McClintock, *Instruments of Statecraft*, 117–120.

72. US Army, "Counter-Guerrilla Operations in the Philippines, 1946–1953," Transcripts of a Seminar on the Huk Champaign (Fort Bragg, NC, June 15, 1961, 61. https://murdercube.com/files/Combined%20Arms/counter guerrillaops.pdf

73. Ibid., 68.

74. Kerkvliet, *The Huk Rebellion*, 102; McClintock, *Instruments of Statecraft*, 123.

75. Walden Bello, ed., *Creating the Third Force: US-Sponsored Low Intensity Conflict in the Philippines*. Institute for Food and Development Policy, 1987. https://foodfirst.org/wp-content/uploads/1987/12/DR02-U.S.-Sponsored-Low-Intensity-Conflict-in-the-Philippines.pdf; Alfred McCoy, *Policing America's Empire: The United States, the Philippines, and the Rise of the Surveillance State* (Madison, WI: University of Wisconsin Press, 2009); Eva-Lotta Hedman, "State of Siege: Political Violence and Vigilante Mobilization in the Philippines," in Bruce Campbell and Arthur Brenner, eds., *Death Squads in Global Perspective* (New York: Palgrave Macmillan, 2000), 125–151; Ramsey Clark, *Right-Wing Vigilantes and US. Involvement: Report of a U.S.-Philippine Fact-Finding Mission to the Philippines* (Manila: Philippine Alliance of Human Rights Advocates, 1987).

76. Jonas, *The Battle for Guatemala*, 288.; Susanne Jonas, "Dangerous Liaisons: The US in Guatemala," *Foreign Policy* (1996), 144–160; McClintock, *The American Connection, Volume 2: State Terror and Popular Resistance in Guatemala*.

77. Kate Doyle, and Carlos Osorio. *US Policy in Guatemala, 1966–1996*. National Security Archive Electronic Briefing Book No. 11, National Security Archive, n.d.; Doyle, Kate, *The Guatemalan Military: What the US*

Files Reveal. National Security Archive Electronic Briefing Book No. 32, National Security Archive, 2000; Douglas Farah, "Papers Show U.S. Role in Guatemalan Abuses," *Washington Post Foreign*, March 11, 1999.

78. Document 5, Doyle and Osorio, *US Policy in Guatemala, 1966–1996*.
79. Viron Vaky, "Memorandum from Viron P. Vaky of the Policy Planning Council to the Assistant Secretary of State for Inter-American Affairs (Oliver), Foreign Relations of the United States, 1964–1968," Volume XXXI, South and Central America; Mexico (Washington, DC: US Office of the Historian, March 29, 1968). https://history.state.gov/historical documents/frus1964-68v31/d102
80. McClintock, *The American Connection, Volume 2: State Terror and Popular Resistance in Guatemala*; Mario Fumerton and Simone Remijnse, "Civil Defense Forces: Peru's CAC and Guatemala's PAC in Comparative Perspective," in Kees Koonings and Dirk Kruijt, eds., *Armed Actors* (London: Zed Books, 2004), 52–72; Simone Remijnse, "Remembering Civil Patrols in Joyabaj, Guatemala," *Bulletin of Latin American Research*, 20, no. 4 (2001), 454–469.
81. Patrick Ball, Paul Kobrak, and Herbert Spirer, *State Violence in Guatemala 1960–1996* (Washington, DC: American Association for the Advancement of Science, 1999).
82. Fumerton and Remijnse, *Civil Defense Forces*, 55.
83. Guatemalan Commission for Historical Clarification (CEH), *Guatemala: Memory of Silence*, American Association for the Advancement of Science (1997).
84. Alan Nairn, "Behind the Death Squads," *The Progressive* (May, 1984), 20–28; Michael McClintock, *The American Connection, Volume 1: State Terror and Popular Resistance in El Salvador* (London: Zed Books, 1985); William Stanley, *The Protection Racket: State Elite Politics, Military Extortion, and Civil War in El Salvador* (Philadelphia, PA: Temple University Press, 1996); Raymond Bonner, *Weakness and Deceit: US Policy in El Salvador* (New York: Times Books, 1984); Russell Crandall, *The Salvador Option* (Cambridge: Cambridge University Press, 2016).
85. David Kirsch, "Death Squads in El Salvador: A Pattern of US Complicity," *Covert Action Quarterly* (Summer 1990), 24.
86. As quoted in Cynthia J. Arnson, "Window on the Past: A Declassified History of Death Squads in El Salvador," in Campbell and Brenner, eds., *Death Squads in Global Perspective*, 93. See Crandall, *The Salvador Option*; Nairn, *Behind the Death Squads*.
87. McClintock, *The American Connection, Volume 1: State Terror and Popular Resistance in El Salvador*. 204, 209–212.
88. Ibid., 212.
89. Kirsch, *Death Squads in El Salvador*.
90. Arnson, *Window on the Past*, 93; Kirsch, *Death Squads in El Salvador*; Leigh Binford, *The El Mozote Massacre* (Tuscon, AZ: University of Arizona Press, 1996); US Bureau of Citizenship and Immigration Services, *El Salvador: The role of ORDEN in the El Salvadoran Civil War*, October 16, 2000, SLV01001.ZAR. www.refworld.org/docid/3dee04524.html

(accessed October 28, 2017); Americas Watch, *El Salvador's Decade of Terror: Human Rights Since the Assassination of Archbishop Romero* (New Haven, CT: Yale University Press, 1991); Crandall, *The Salvador Option*.

91. McClintock, *The American Connection, Volume 1: State Terror and Popular Resistance in El Salvador*, 34; William Leo Grande and Carla Ann Robbins, "Oligarchs and Officers: The Crisis in El Salvador," *Foreign Affairs*, 58, no. 5 (1980), 1084–1103; Stanley, *The Protection Racket: State Elite Politics*; Commission on the Truth for El Salvador, *From Madness to Hope: The 12-year War in El Salvador: Report of the Commission on the Truth for El Salvador*, UN Security Council, March 15, 1993. www.usip.org/sites/default/files/file/ElSalvador-Report.pdf

92. As quoted in Arnson, *Window on the Past*, 91.

93. McClintock, *The American Connection, Volume 1: State Terror and Popular Resistance in El Salvador*, 34–36; Arnson, *Window on the Past*, 92.

94. Arnson, *Window on the Past*, 111.

95. Doug Stokes, *America's Other War: Terrorizing Colombia* (New York: Zed Books, 2005); Doug Stokes, "Why the End of the Cold War Doesn't Matter: The US War of Terror in Colombia," *Review of International Studies*, 29 (2003), 569–585; Victoria Sanford, "Learning to Kill by Proxy: Colombian Paramilitaries and the Legacy of Central American Death Squads, Contras, and Civil Patrols," *Social Justice* (2003); Michael Lopez, "The U.S. and its Responsibility for Counter-Insurgency Operations in Colombia," *Colombia Bulletin*, Summer (1998); Jasmin Hristov, *Blood and Capital: The Paramilitarization of Colombia* (Toronto: Between the Lines, 2009); Jasmin Hristov, *Paramilitarism and Neoliberalism: Violent Systems of Capital Accumulation in Colombia and Beyond* (London: Pluto Press, 2014); Human Rights Watch, "Colombia's Killer Networks: The Military-Paramilitary Partnership and the United States," *Human Rights Watch* (November, 1996).

96. As quoted in McClintock, *Instruments of Statecraft*, 222.

97. Dennis Rempe, "Guerrillas, Bandits, and Independent Republics: US Counterinsurgency Efforts in Colombia 1959–1965," *Small Wars and Insurgencies*, 6, no. 3 (1995), 304–327; Dennis Rempe, "The Past as Prologue? A History of U.S. Counterinsurgency Policy in Colombia, 1958–66," *Strategic Studies Institute Monograph* (March, 2002).

98. Lopez, "The U.S. and its Responsibility for Counter-Insurgency Operations in Colombia."

99. Javier Giraldo, "Genocidio En Colombia: Tipicidad y Estrategia," *Desde Los Margenes*, 2004.

100. As quoted in Human Rights Watch, "Colombia's Killer Networks: The Military-Paramilitary Partnership and the United States," *Human Rights Watch* (November, 1996); see also Human Rights Watch, "The 'Sixth Division': Military-Paramilitary Ties and U.S. Policy in Colombia," *Human Rights Watch* (2001); Giraldo, *Genocidio En Colombia*.

101. See McClintock, *Instruments of Statecraft*, 252.

102. Human Rights Watch, "The History of the Military-Paramilitary Partnership," In *Colombia's Killer Networks* (New York: Human Rights Watch, 1996).

103. Javier Giraldo, *Colombia: The Genocidal Democracy* (Monroe, ME: Common Courage Press, 1996), 79.
104. As quoted in Giraldo, *Colombia: The Genocidal Democracy*, 80.
105. Anon., *Plans to Counteract Communism and Castro Penetration in Colombia and Other Parts of Latin America*. Memorandum of Conversation. Department of State, 1960. www.icdc.com/~paulwolf/colombia/gerberich15dec1960.htm (accessed July 20, 2012).
106. Hristov, *Blood and Capital*; Hristov, *Paramilitarism and Neoliberalism*; David Maher, "The Fatal Attraction of Civil War Economies: Foreign Direct Investment and Political Violence. A Case Study of Colombia," *International Studies Review*, 17, no. 2 (2015), 217–248; David Maher and Andrew Thomson, "The Terror that Underpins the 'Peace': The Political Economy of Colombia's Paramilitary Demobilisation Process," *Critical Studies on Terrorism*, 4, no. 1 (2011), 95–113; Edwin Cruz, "Los estudios sobre el paramilitarismo en Colombia," *Analysis Politico*, 60 (May–August, 2007), 117–134.
107. Mike Gravel, *The Pentagon Papers*, Vol. 2 (Boston, MA: Beacon Press, 1971), 39; Douglas Valentine, *The Phoenix Program* (Lincoln, NE: iUniverse.com, Inc., 2000), 49.
108. Roger Hilsman, Personal Papers. Countries Files, 1962a. Vietnam: "A Strategic Concept for South Vietnam," February 2, 1962. RHPP-003-006. John F. Kennedy Presidential Library and Museum www.jfklibrary.org/Asset-Viewer/Archives/RHPP-003-006.aspx; Roger Hilsman, "Memorandum From the Director of the Bureau of Intelligence and Research (Hilsman) to the Assistant Secretary of State for Far Eastern Affairs (Harriman)," Washington, DC, December 19, 1962. Kennedy Library, National Security Files, Vietnam Country Series https://history.state.gov/historicaldocuments/frus1961-63v02/d332; Milton Osborne, *Strategic Hamlets in South Viet-nam: A Survey and Comparison* (Ithaca, NY: Cornell University Press, 1965).
109. Colonel Francis Kelly, *Vietnam Studies: US Army Special Forces 1961–1971* (Washington, DC: Department of the Army, 1985), 6.
110. Richard Shultz, "The Vietnamization-Pacification Strategy of 1969–1972: A Quantitative and Qualitative Reassessment, " in Richard Hunt and R. Shultz, eds., *Lessons from an Unconventional War: Reassessing U.S. Strategies for Future Conflicts* (New York: Pergamon Press, 1982), 56.
111. As quoted in McClintock, *Instruments of Statecraft*, 24.
112. National Security Council, National Security Action Memorandum (NSAM 56), 1961. "National Security Action Memorandum Number 56: Evaluation of Paramilitary Requirements," June 28, 1961 Kennedy Library, National Security Council www.jfklibrary.org/Asset-Viewer/ex2GImrvWU-Z9Zqyga7ryQ.aspx
113. As quoted in Christopher Ives, *US Special Forces and Counterinsurgency in Vietnam: Military Innovation and Institutional Failure, 1961–1963* (New York: Routledge, 2007), 103, also see p. 24.
114. James Dunigan and Albert Nofi, *Dirty Little Secrets of the Vietnam War* (New York: St. Martin's Press, 1999).

115. John Prados, *Presidents' Secret Wars* (Chicago, IL: Elephant Paperbacks, 1996), 252; John Prados, *The Hidden History of the Vietnam War* (Chicago, IL: Ivan R. Dee, 1995).

116. Valentine, *The Phoenix Program*, 277.

117. Valentine, *The Phoenix Program*, 162–163, 166–167, 170.

118. Joe F. Leeker, *The History of Air America* (Dallas, TX: University of Texas Press, 2015, second ed.

119. Valentine, *The Phoenix Program*, 289.

120. Colonel Andrew Finlayson, "A Retrospective on Counterinsurgency Operations," Homepage of CIA, 2008. www.cia.gov/library/center-for-the-study-of-intelligence/csi-publications/csi-studies/studies/vol51no2/a-retrospective-on-counterinsurgency-operations.html (accessed December 21, 2016).

121. Goergie Anne Geyer, "The CIA's Hired Killers," *True Magazine* (February, 1970).

122. Valentine, *The Phoenix Program*, 315.

123. Dunigan and Nofi, *Dirty Little Secrets of the Vietnam War*, 181.

124. Valentine, *The Phoenix Program*, 160.

125. Robert M Blackburn, *Mercenaries and Lyndon Johnson's "More Flags": The Hiring of Korean, Filipino and Thai Soldiers in the Vietnam War* (Jefferson, NC: McFarland, 1994).

126. Laura A. Dickinson, *Outsourcing War and Peace* (New Haven, CT: Yale University Press, 2011), 29.

127. Christopher Robbins, *Air America* (New York: Avon Books, 1979 [1990]), 37–39.

128. William M. Leary, "CIA Air Operations in Laos 1955–1974: Supporting the Secret War," CIA Webpage, 2007 (last updated 2008). www.cia.gov/library/center-for-the-study-of-intelligence/csi-publications/csi-studies/studies/winter99-00/art7.html (accessed September 4, 2016); William. M. Leary, *Perilous Missions: Civil Air Transport and CIA Covert Operations in Asia* (Tuscaloosa, AL: University of Alabama Press, 1984, second ed.).

129. Valentine, *The Phoenix Program*, 31–33.

130. Leeker, *The History of Air America*.

131. Lieutenant General Joseph M. Heiser, Jr. *Logistic Support*. Vietnam Studies (Washington, DC: Department of the Army, 1974), 88.

132. Allison Stanger, *One Nation Under Contract: The Outsourcing of American Power and the Future of Foreign Policy* (London: Yale University Press, 2009), 84.

133. As quoted in Stephen J. Zamparelli, "Contractors in the Battlefield: What Have We Signed Up For?" *Air Force Journal of Logistics*, 23 (Fall, 1999), 10.

134. Heiser, *Logistic Support*, 4.

135. William Hartung, "Saudi Arabia: Mercenaries, Inc.," *The Progressive* (April 1, 1996).

136. Lieutenant Colonel Carroll Dunn, *Base Development in South Vietnam 1965–1970*. (Washington, DC: US Department of the Army, 1991 [1972]); Pratap Chatterjee, *Halliburton's Army*. (New York: Nation Books, 2009), 25–28; Deborah Kidwell, "Public War, Private Fight? The United States

and Private Military Companies," Global War on Terrorism Occasional Paper 12 (Fort Leavenworth, KS: Combat Studies Institute Press, 2005).

137. Robbins, *Air America*; Leeker, *The History of Air America*; James E. Parker, *Covert Ops: The CIA's Secret War in Laos*. (St. Martin's Press, 1997).

138. Leeker, *The History of Air America*; Robbins, *Air America*, 144.

139. As quoted in Hartung, *Saudi Arabia: Mercenaries, Inc.*

140. Mark Moyar, *Phoenix and the Birds of Prey: Counterinsurgency and Counterterrorism in Vietnam* (Lincoln, NB: University of Nebraska Press, 2007), 164; Valentine, *The Phoenix Program*, 175, 212, 258, 271, 358.

141. Valentine, *The Phoenix Program*, 353.

142. Valentine, *The Phoenix Program*, 397, 405, 415; Moyar, *Phoenix and the Birds of Prey*; Robbins, *Air America*; Heiser, *Logistic Support*.

143. Valentine, *The Phoenix Program*, 415.

144. Leeker, *The History of Air America*.

145. Robbins, *Air America*, 145.

146. Robbins, *Air America*, 145; Leeker, *The History of Air America*; "AA in South Vietnam," Volume II, 35.

147. Heiser, *Logistic Support*, 13, 84.

148. Kidwell, *Public War, Private Fight?*

149. Victor Anthony and Richard Sexton, *The United States Airforce in South East Asia: The War in Northern Laos 1954–1973* (Washington, DC: Center for Air Force History, US Air Force, 1993), 42–43, 379. http://nsarchive.gwu.edu/NSAEBB/NSAEBB248/war_in_northern_laos.pdf; also see John Prados, *Fighting the War in South East Asia, 1961–1973*, National Security Archive Electronic Briefing Book No. 248 (Washington, DC: National Security Archive, April 9, 2008). http://nsarchive.gwu.edu/NSAEBB/NSAEBB248/

150. Also see Carl W. Borklund, *The Department of Defense* (New York: Praeger, 1968), 74–75, 83–84.

151. Dickinson, *Outsourcing War and Peace*, 30.

152. Kidwell, *Public War, Private Fight?*, 21.

153. Hartung, *Saudi Arabia: Mercenaries, Inc.*

4. Reagan, Low-Intensity Conflict, and the Expansion of Para-Institutional Statecraft

1. Michael McClintock, *Instruments of Statecraft: U.S. Guerrilla Warfare, Counterinsurgency, and Counterterrorism, 1940–1990* (New York: Pantheon Books, 1992), 335.

2. Fred Halliday, *The Making of the Second Cold War* (London: Verso, 1987); Simon Dalby, *Creating the Second Cold War: The Discourse of Politics* (London: Pinter Publishers, 1990).

3. For a good overview, see Ivan Molloy, *Rolling Back Revolution: The Emergence of Low Intensity Conflict* (London: Pluto Press, 2001). For US military documents which outline low-intensity conflict and a series of conferences and work on it, see Center for Low Intensity Conflict,

"Volume 1 Analytical Review of Low-Intensity Conflict" (Fort Mondroe, VA: United States Army Training and Doctrine Command, August 1, 1986). www.dtic.mil/dtic/tr/fulltext/u2/a185971.pdf; Center for Low Intensity Conflict, "Volume II: Low Intensity Conflict, Issues and Recommendations" (Fort Mondroe, VA: United States Army Training and Doctrine Command, August 1, 1986); Also see Thomas Bodenheimer and Robert Gould, *Rollback: Right Wing Power in US Foreign Policy* (Boston, MA: South End Press, 1989), 5.

4. Bodenheimer and Gould, *Rollback: Right Wing Power in US Foreign Policy*.
5. Ronald Reagan, *National Security Strategy of the United States* (Washington, DC: The White House, 1987), 34.
6. Ibid., 33.
7. James Motley, "A Perspective on Low-Intensity Conflict," *Military Review*, January (1985); Michael T. Klare and Peter Kornbluh, "The New Interventionism," in idem, eds., *Low Intensity Warfare: Counterinsurgency, Proinsurgency and Anti-Terrorism in the Eighties* (New York: Pantheon Books, 1988), 5.
8. US Army, *Military Operations in Low Intensity Warfare: FM 100-20* (Washington, DC: Headquarters, Department of the Army, 1990), 1-1.
9. Geoff Simons, *The Vietnam Syndrome: Impact on US Foreign Policy* (Basingstoke: Macmillan Press, 1998), 288; Molloy, *Rolling Back Revolution*.
10. Molloy, *Rolling Back Revolution*, 2.
11. Klare and Kornbluh, *The New Interventionism*, 3–20.
12. As quoted in Jonathan Marshall, Peter Dale Scott, and Jane Hunter, *The Iran-Contra Connection: Secret Teams and Covert Operations in the Reagan Era* (London: South End Press, 1987), 189.
13. Marshall, Scott and Hunter, *The Iran-Contra Connection*, 8; Greg Grandin, *Empire's Workshop: Latin America, the United States, and the Rise of the New Imperialism* (New York: Owl Books, 2006), 62.
14. Grandin, *Empire's Workshop*, 119.
15. Michael McClintock, *The American Connection, Volume 1: State Terror and Popular Resistance in El Salvador* (London: Zed Books, 1985), 340, 317–325.
16. Bello, Walden. F., ed., *Creating the Third Force: US Sponsored Low Intensity Conflict in the Philippines*. Institute for Food and Development Policy, (1987), 90. https://foodfirst.org/wp-content/uploads/1987/12/DR02-U.S.-Sponsored-Low-Intensity-Conflict-in-the-Philippines.pdf
17. Steven Emerson, *Secret Warriors: Inside the Covert Military Operations of the Reagan Era* (New York: Putnam Publishing Group. 1988).
18. McClintock, *Instruments of Statecraft*, 349; Bodenheimer and Gould, *Rollback: Right Wing Power in US Foreign Policy*, 102. For CIA figures, see William I. Robinson and Kent Norsworthy, *David and Goliath: Washington's War Against Nicaragua* (London: Zed Books, 1987), 35.
19. As cited in McClintock, *Instruments of Statecraft*, 351.
20. Robinson and Norsworthy, *David and Goliath: Washington's War Against Nicaragua*, 337; William Blum, *Rogue State: A Guide to the World's Only Superpower* (London: Zed Books, 2003, second ed.); Kenneth Conboy, *The*

Cambodian Wars: Clashing Armies and CIA Covert Operations (Lawrence, KS: University Press of Kansas, 2013).

21. Select Committee on Foreign Affairs, *Possible Violation Or Circumvention of the Clark Amendment* (Washington, DC: US Government Printing Office, 1987). http://babel.hathitrust.org/cgi/pt?id=pst.000013683399;page=root; view=image;size=100;seq=1

22. George Wright, *The Destruction of a Nation: United States' Policy Towards Angola Since 1945* (London: Pluto Press, 1997), 126–127; Victoria Brittain, *Death of Dignity: Angola's Civil War* (London: Pluto Press, 1998); Sean Cleary, "Angola: A Case Study of Private Military Involvement," in Jakkie Cilliers and Peggy Mason, eds., *Peace, Profit or Plunder* (Johannesburg: Institute for Security Studies, 1999), 141–174.

23. Steve Galster, "Afghanistan the Making of US Policy 1973–1990: Volume II: Afghanistan Lessons from the Last War," National Security Archive, October 9, 2001. http://nsarchive.gwu.edu/NSAEBB/NSAEBB57/essay.html; John Prados, "Notes on the CIA's Secret War in Afghanistan," *Journal of American History*, 89, no. 2 (2002), 466–471; John Prados, *Presidents' Secret Wars* (Chicago, IL: Elephant Paperbacks, 1996), 361.

24. Michael McClintock, *The American Connection, Vol. 1: State Terror and Popular Resistance in El Salvador*, 333, 337, 342.

25. Michael McClintock, *The American Connection, Vol. 2: State Terror and Popular Resistance in Guatemala* (London: Zed Books, 1985), 249–252; Guatemalan Commission for Historical Clarification (CEH), *Guatemala: Memory of Silence*, American Association for the Advancement of Science, 1997. www.aaas.org/sites/default/files/migrate/uploads/mos_en.pdf

26. Molloy, *Rolling Back Revolution*, 136, 137–140; Bello, *Creating the Third Force*, 90–100; David Kowalewski, "Cultism, Insurgency, and Vigilantism in the Philippines," *Sociology of Religion*, 52, no. 3 (1991), 241–253; David Kowalewski, "Counterinsurgent Paramilitarism: A Philippine Case Study," *Journal of Peace Research*, 29, no. 1 (Feb., 1992), 71–84; Adele Oltman and Dennis Bernstein, "The El Salvador of the Pacific: Counterinsurgency in the Philippines," *Covert Action Information Bulletin*, no. 29 (Winter, 1988).

27. Yelena Biberman, "Self-Defense Militias, Death Squads, and State Out-sourcing of Violence in India and Turkey," *Journal of Strategic Studies* (2016), 1–31; Ceren Belge, "State Building and the Limits of Legibility: Kinship Networks and Kurdish Resistance in Turkey," *International Journal of Middle East Studies*, 43, no. 1 (2011), 95–114; Serdar Celik, *Turkey's Killing Machine: The Contra-Guerrilla Force Kurdistan*, Report No. 17 February/March 1994. www.hartford-hwp.com/archives/51/017.html

28. Ronald Reagan, "Address Before a Joint Session of the Congress on the State of the Union," February 6, 1985. www.presidency.ucsb.edu/ws/?pid=38069; Ronald Reagan, "Radio Address to the Nation on Foreign Policy Achieve-ments," 1988. www.presidency.ucsb.edu/ws/index.php?pid=36300

29. Ronald Reagan, "Seventh State of the Union Address," 1988. www.presidency.ucsb.edu/ws/?pid=36035

30. Prados, *Presidents' Secret Wars*, 357.

31. John H. Marsh, "Keynote Address by the Honorable John O. Marsh Jr.," in Frank Barnett, Hugh Tovar, and Richard Shultz, eds., *Special Operations in US Strategy* (New York: National Defense University Press, 1984), 17–27.

32. Sara Diamond, *Roads to Dominion: Right-Wing Movements and Political Power in the United States* (New York: Guildford Press, 1995), 240.

33. Diamond, *Roads to Dominion*, 218; Robinson and Norsworthy, *David and Goliath: Washington's War Against Nicaragua*, 92.

34. Marshall, Scott, and Hunter, *The Iran-Contra Connection*; Right Web, "World-Anti Communist League," *Right Web*. http://rightweb.irc-online. org/articles/display/World_Anti-Communist_League#P10671_2152317 (accessed November 29, 2011).

35. Russ Bellant, *Old Nazis, the New Right, and the Republican Party* (New York: South End Press, 1991), 83–88.

36. Diamond, *Roads to Dominion*, 223; Bodenheimer and Gould, *Rollback: Right Wing Power in US Foreign Policy*.

37. Diamond, *Roads to Dominion*; Edward Herman and Gerry O'Sullivan, *The "Terrorism" Industry: The Experts and Institutions That Shape Our View of Terror* (New York: Pantheon, 1989).

38. Marshall, Scott, and Hunter, *The Iran-Contra Connection*.

39. Bodenheimer and Gould, *Rollback: Right Wing Power in US Foreign Policy*, 54.

40. Diamond, *Roads to Dominion*, 121, 207, 216, 228, 243; Herman and O'Sullivan, *The "Terrorism" Industry*.

41. Robert Parry, "How Reagan's Propaganda Succeeded," *Consortium News*, 2015[2010].https://consortiumnews.com/2015/06/04/how-reagans-propaganda-succeeded/; Walter Raymond, "Memorandum for William Clark: Democracy Initiative and Information Programs," Memorandum National Security Council, November 4, 1982. https://consortiumnews.com/wp-content/uploads/2016/05/Nov482.pdf

42. Edward S. Herman, *The Real Terror Network: Terrorism in Fact and Propaganda* (Boston, MA: South End Press, 1982); John Gastil, "Undemocratic Discourse: A Review of Theory and Research on Political Discourse," *Discourse & Society*, 3, no. 4 (1992), 469–500; J. Wilson Gibson, "Paramilitary Fantasy Culture and the Cosmogonic Mythology of Primeval Chaos and Order," *Vietnam Generation*, 1, no. 3 (1989).

43. Richard Schultz, "Low Intensity Conflict," in Stuart M. Butler et al., eds., *Mandate for Leadership II: Continuing the Conservative Revolution* (Washington, DC: Heritage Foundation, 1984), 264–270, 267–268; The Committee of Santa Fe, *A New Inter-American Policy for the Eighties*, prepared by the Committee of Santa Fe for the Council for Inter-American Security, Inc. (Washington, DC: Council for Inter-American Security, 1989); The Heritage Foundation, "Mandate for Leadership: Policy Management in a Conservative Administration," Charles L Heatherley, ed., January (1981).

44. Diamond, *Roads to Dominion*, 134, 205, 210, 211.

45. Ibid., 207.

46. Ibid., 222–223, 224, 240; Walden Bello, "Counterinsurgency's Proving Ground: Low-Intensity Warfare in the Philippines," in Michael T. Klare and Peter Kornbluh, eds., *Low Intensity Warfare: Counterinsurgency, Pro-insurgency and Anti-Terrorism in the Eighties* (New York: Pantheon Books, 1988), 175–178.

47. Select Committee to Investigate Covert Arms Transactions with Iran Report of the Congressional Committees Investigating the Iran-Contra Affair Washington D.C.: Senate Report No. 216, November 17, 1987. https://ia802205.us.archive.org/16/items/reportofcongress87unit/reportofcongress87unit.pdf; Bob Woodward, *Veil: The Secret Wars of the CIA, 1981–1987* (New York: Simon and Schuster, 2005[1987]), 434–489.

48. They complete extensive research on the emergence of the private security industry in the 1980s: Herman and O'Sullivan, *The "Terrorism" Industry*, 51.

49. Prados, *Presidents' Secret Wars*, 497, 498–499.

50. Clyde Farnsworth, "'The Company' as Big Business," *New York Times*, January 4, 1987.

51. CIA, "Liquidation Plan" CIA Reading Room, 1976. www.cia.gov/library/readingroom/docs/197616.pdf, 3; Joe F. Leeker, *The History of Air America* (Dallas, TX: University of Texas, 2015, second ed.); Robbins, *Air America*, 283–309.

52. Robbins, *Air America*, 283–309.

53. CIA, "Liquidation Plan," 3.

54. Prados, *Presidents' Secret Wars*, 497–498.

55. Summit Aviation, 2017. http://summit-aviation.com/about/.

56. Robert Block, "Airline Swaps Gun-Running for Good Works: Southern Air Transport was Run by the CIA and Started the Iran-Contra Scandal. But Now, Its Staff Assure Robert Block, Its Main Job is Saving Somalis," *The Independent*, December 24, 1992. www.independent.co.uk/news/world/airline-swaps-gun-running-for-good-works-southern-air-transport-was-run-by-the-cia-and-started-the-1565324.html; Prados, *Presidents' Secret Wars*, 498.

57. DynCorp International, "A Brief History of DynCorp International," *DynCorp International Website*, 2012. www.dyn-intl.com/about-us/history.aspx; David Isenberg, *Shadow Force: Private Security Contractors in Iraq* (Westport, CT: Praeger, 2009), 2.

58. Funding Universe, n.d., *Dyncorp History Funding Universe Website*. www.fundinguniverse.com/company-histories/dyncorp-history/

59. Greg Guma, "The CIA, DynCorp, and the Shoot Down in Peru," *Zmag Online*, December 2009. www.zcommunications.org/the-cia-dyncorp-and-the-shoot-down-in-peru-by-greg-guma; Alvear Restrepo, "Private Security Transnational Enterprises in Colombia: Case Study Plan Colombia," *Corporación Colectivo De Abogados* (February, 2008).

60. Jeremy Bigwood, "DynCorp in Colombia: Outsourcing the Drug War," *Corpwatch*, 2001. http://corpwatch.org/article.php?id=672

61. Lieutenant General Joseph M. Heiser, Jr., *Logistic Support*. Vietnam Studies (Washington, DC: Department of the Army, 1974).

62. Laura Dickinson, *Outsourcing War and Peace* (New Haven, CT: Yale University Press, 2011), 29.

63. James Jay Carafano, *Private Sector, Public War* (Westport, CT: Praeger, 2008).

64. Christopher Robbins, *Air America* (New York: Avon Books, 1979 [1990]), 283–285.

65. For more see Christopher Kinsey, *Corporate Soldiers and International Security: The Rise of Private Military Companies* (London: Routledge, 2007), 98–101.

66. Sourcewatch, "MPRI, Inc." 2010. www.sourcewatch.org/index.php?title= Military_Professional_Resources_Inc.; Isenberg, *Shadow Force: Private Security Contractors in Iraq.*

67. Herman and O'Sullivan, *The "Terrorism" Industry*, 123.

68. Juan Tamayo, "Colombia: Private Firms Take on US Military Role in Drug War," *Miami Herald*, May 22, 2001. www.corpwatch.org/article.php?id=11094; Alan Axelrod, *Mercenaries: A Guide to Private Armies and Private Military Companies* (London: CQ Press, 2014).

69. As quoted in Tom Burghardt, "'Managing' Data and Dissent," *Pacific Free Press*, April 4, 2010.

70. Ken Guggenheim, "Drug Fight in Colombia Questioned," *Washington Post*, June 5, 2001; Peter Dale Scott, *Drugs, Oil, and War: The United States in Afghanistan, Colombia, and Indochina* (Lanham, MD and Oxford: Rowman & Littlefield 2003).

71. Prados, *Presidents' Secret Wars*, 498.

72. Scott Shane, Stephen Grey, and Margot Williams, "C.I.A. Expanding Terror Battle Under Guise of Charter Flights," *New York Times*, May 31, 2005; Trevor Paglen and Adam Clay Thompson, *Torture Taxi: On the Trail of the CIA's Rendition Flights* (Cambridge: Icon Books, 2007).

73. Pratap Chatterjee, *Halliburton's Army* (New York: Nation Books, 2009); Bruce E. Stanley, *Outsourcing Security: Private Military Contractors and US Foreign Policy* (Lincoln, NB: University of Nebraska Press, 2015), 51–54.

74. Stanley, *Outsourcing Security*, 38.

75. David M. Kotz, *The Rise and Fall of Neoliberal Capitalism* (Cambridge, MA: Harvard University Press, 2015), 18, 46, 72; John Williamson, *What Washington Means by Policy Reform*, 1990. www.iie.com/publications/ papers/paper.cfm?ResearchID=486

76. OMB Circular No. A-76 1983 (rev.), 48 Fed. Reg. 37,110, 37,114 4a (August 16, 1983): LTC William Beatty, *Department of Defense, Inc. Are We Ready to Become an Extension of Corporate America?* (Carlisle Barracks, PA: Strategy Research Project, US Army War College, 1999). www.dtic.mil/ dtic/tr/fulltext/u2/a362978.pdf (accessed October 12, 2017).

77. Michal Laurie Tingle, "Privatization and the Reagan Administration: Ideology and Application," *Yale Law & Policy Review*, 6, no. 1, Article 12 (1988), 233; Beatty, *Department of Defense, Inc.*

78. Elke Krahmann, *States, Citizens, and the Privatization of Security* (Cambridge: Cambridge University Press, 2010), 76.

79. Chatterjee, *Halliburton's Army*, 52.

80. Deborah Kidwell, *Public War, Private Fight? The United States and Private Military Companies* (Fort Leavenworth, KS: Combat Studies Institute Press,[2005]), 18.
81. Kidwell, *Public War, Private Fight?*; Paul Starr, "The Limits of Privatization," *Proceedings of the Academy of Political Science*, 36, no. 3 (1987), 124–137.
82. Deborah Avant, *The Market for Force: The Consequences of Privatising Security* (Cambridge: Cambridge University Press, 2005), 36.
83. Beth Bailey, "The Army in the Marketplace: Recruiting an All-Volunteer Force," *Journal of American History*, 94, no. 1 (June, 2007), 47–74; Patricia M. Shields and Kay Hofer, "Consequences of privatization," *The Bureaucrat*, 17, no. 4 (1988), 33–37; Bernard Rostker, *I Want You! The Evolution of the All Volunteer Force* (Santa Monica, CA: RAND, 2006); Molly Clever and David R. Segal, "After Conscription: The United States and the All-Volunteer Force," *Sicherheit Und Frieden (S F) / Security and Peace*, 30, no. 1 (2012), 9–18.
84. Maya Eichler, "Private Security and Gender," in Rita Abrahamsen and Anna Leander, eds., *Routledge Handbook of Private Security Studies* (London: Routledge, 2016), 161; Krahmann, *States, Citizens, and the Privatization of Security*.
85. Marshall, Scott, and Hunter, *The Iran-Contra Connection*, 7.
86. Robinson and Norsworthy, *David and Goliath: Washington's War Against Nicaragua*, 82–83.
87. Holly Sklar, *Washington's War on Nicaragua* (Cambridge, MA: South End Press, 1988), 37.
88. Gabriel Kolko, *Confronting the Third World: United States Foreign Policy 1945–1980* (New York: Pantheon Books, 1988), 288.
89. Grace Livingstone, *America's Backyard* (London: Zed Books, 2009), 77; Robinson and Norsworthy, *David and Goliath: Washington's War Against Nicaragua*, 46; Select Committees, US House of Representatives and Senate, *Report of the Congressional Committees Investigating the Iran-Contra Affair* (Washington, DC: US House of Representatives and US Senate, 1987).
90. Sklar, *Washington's War on Nicaragua*, 100.
91. Select Committees, US House of Representatives and Senate, *Report … Investigating the Iran-Contra Affair*, 339; Also see Robinson and Norsworthy, *David and Goliath: Washington's War Against Nicaragua*, 49.
92. CIA, *Psychological Operations in Guerrilla Warfare* (Langley, VA: CIA, 1983). www.cia.gov/library/readingroom/docs/CIA-RDP86M00886R001300010029-9.pdf; Robinson and Norsworthy, *David and Goliath: Washington's War Against Nicaragua*, 46.
93. Livingstone, *America's Backyard*; Grandin, *Empire's Workshop*, 90.
94. Grandin, *Empire's Workshop*, 116.; Peter Kornbluh, "Nicaragua: US Proinsurgency Warfare Against the Sandanistas," in Michael T. Klare and P. Kornbluh, eds., *Low Intensity Warfare* (New York: Pantheon Books, 1988), 142–143; David Rogers and David Ignatius, "The Contra Fight," *Wall Street Journal*, March 6, 1985.
95. Kornbluh, "Nicaragua: US Proinsurgency Warfare Against the Sandinistas," 142.

96. Select Committees, US House of Representatives and Senate, *Report ... Investigating the Iran-Contra Affair*, 648.

97. Robinson and Norsworthy, *David and Goliath: Washington's War Against Nicaragua*, 88.

98. Grandin, *Empire's Workshop*, 144.

99. Ken Silverstein, *Private Warriors* (London: Verso, 2000), 88.

100. Robinson and Norsworthy, *David and Goliath: Washington's War Against Nicaragua*, 185.

101. Select Committees, US House of Representatives and Senate, *Report ... Investigating the Iran-Contra Affair*, 124; Sklar, *Washington's War on Nicaragua*, 276.

102. Oliver North, quoted in Select Committees, US House of Representatives and Senate, *Report ... Investigating the Iran-Contra Affair*, 338.

103. Herman and O'Sullivan, *The "Terrorism" Industry*, 134.

104. Quoted in Sklar, *Washington's War on Nicaragua*, 276.

105. Select Committees, US House of Representatives and Senate, *Report ... Investigating the Iran-Contra Affair*, 124.

106. The Boland Amendment is the title given to a series of three legislations prohibiting military or "lethal" aid to the Contras in the explicit quest to overthrow the Sandinista government.

107. Robinson and Norsworthy, *David and Goliath: Washington's War Against Nicaragua*, 88–89.

108. Select Committees, US House of Representatives and Senate, *Report ... Investigating the Iran-Contra Affair*, 327.

109. Sklar, *Washington's War on Nicaragua*, 258. Also see Robinson and Norsworthy, *David and Goliath: Washington's War Against Nicaragua*.

110. Marshall, Scott, and Hunter, *The Iran-Contra Connection*, 217, 341.

111. Select Committees, US House of Representatives and Senate, *Report ... Investigating the Iran-Contra Affair*, 67, 338.

112. Quoted in Sklar, *Washington's War on Nicaragua*, 279.

113. Select Committees, US House of Representatives and Senate, *Report ... Investigating the Iran-Contra Affair*.

114. Marshall, Scott, and Hunter, *The Iran-Contra Connection*, 7.

115. Bodenheimer and Gould, *Rollback: Right Wing Power in US Foreign Policy*, 110.

116. Quoted in Steven Metz, "Rethinking Insurgency," *Strategic Studies Institute* (2007), 5.

5.   Continuity After the Cold War and the Consolidation of Para-Institutional Complexes

1. Vassilis Fouskas and Bülent Gökay, *The New American Imperialism: Bush's War on Terror and Blood for Oil* (Westport, CT: Greenwood Publishing Group, 2005), 20–23.

2. Francis Fukuyama, "The End of History?" *The National Interest*, 16 (1989), 3–18.

3. Cyrus Bina and Behzad Yaghmaian, "Post-War Global Accumulation and the Transnationalisation of Capital," *Capital & Class*, 15, no. 1 (1991), 107–130: William I. Robinson, *A Theory of Global Capitalism: Production, Class and State in a Transnational World* (Baltimore, MD: John Hopkins University Press, 2004); William I. Robinson, *A Theory of Global Capitalism: Production, Class and State in a Transnational World* (Baltimore, MD: Johns Hopkins University Press, 2004), 78.

4. Christopher Layne, *The Peace of Illusions: American Grand Strategy from 1940 to the Present* (Ithaca, NY: Cornell University Press, 2006); Doug Stokes and Sam Raphael, *Global Energy Security and American Hegemony* (Baltimore, MD: John Hopkins University Press, 2010); Andrew Bacevich, *American Empire* (London: Harvard University Press, 2002); Noam Chomsky, *Year 501: The Conquest Continues* (London: Verso Press, 1993); Doug Stokes, "The Heart of Empire?: Theorising US Empire in an Era of Transnational Capitalism," *Third World Quarterly*, 26, no. 2 (2005), 217–236.

5. Dick Cheney, *Annual Report of the Secretary of Defense to the President and to the Congress* (Washington, DC: Department of Defense, 1990). http://osdhistory.defense.gov/docs/1990%20DoD%20Annual%20 Report%20-%20Cheney.pdf

6. General A. M. Gray, "Defense Policy for the 1990s," *Marine Corps Gazette*, May (1990), 18; see also John Quigley, *The Ruses for War* (New York: Prometheus Books, 2007), 403.

7. George Bush, *President George H. W. Bush Address to the Congress 6 March 1991*, Washington, DC, March 6, 1991.

8. George Bush, *National Security Strategy of the United States August 1991*, Washington, DC, August, 1991; emphasis added.

9. Ibid.

10. John Shalikashvili, *National Military Strategy* (Washington, DC: Joint Chiefs of Staff, Department of Defense, 1995).

11. Jochen Hippler, "Counterinsurgency and Political Control," *INEF Report*, 81 (2006).

12. Beau Grosscup, "The American Doctrine of Low Intensity Conflict in the New World Order," in A. Grammy and K. Bragg, eds., *United States Third World Relations in the New World Order* (New York: Nova Publishers, 1996), 57.

13. Lora Lumpe, "US Foreign Military Training: Global Reach, Global Power, and Oversight Issues," *Foreign Policy in Focus* (May, 2002).

14. Thomas Ferguson, "Preface: Rethinking the State and 'Free Markets' in Neoliberalism," in Ronald Cox, ed., *Corporate Power and Globalization in US Foreign Policy* (New York: Routledge, 2012), xi; William Blum, *Rogue State: A Guide to the World's Only Superpower* (London: Zed Books, 2003, second ed.).

15. US Army, *Foreign Internal Defense: Tactics, Techniques, and Procedures for Special Forces FM 31-20-3* (Washington, DC: Department of the Army, 1994), G-1.

16. Ibid., 1-6.

17. William S. Cohen, *Annual Report to the President and the Congress* (Washington, DC: Department of Defense, 1997).

18. US Army, *Doctrine for Special Forces Operations FM 31-20* (Washington, DC: Department of the Army, 1990), 1-11, 1-12.

19. US Army. *Foreign Internal Defense: Tactics, Techniques, and Procedures for Special Forces FM 31-20-3* (Washington, DC: Department of the Army, 1994), 1-19. See also US Army. *Operations in a Low Intensity Conflict: FM 7-98* (Washington, DC: Department of the Army, 1992), 10-5.

20. US Army, *Doctrine for Special Forces Operations FM 31-20* (Washington, DC: Department of the Army, 1990), 8-5; US Army, *Military Operations in Low Intensity Warfare: FM 100-20* (Washington, DC: Headquarters, Department of the Army, 1990).

21. US Army, *Foreign Internal Defense: Tactics, Techniques, and Procedures for Special Forces FM 31-20-3* (Washington, DC: Department of the Army, 1994).

22. For example, Govinda Clayton and Andrew Thomson, "Civilianizing Civil Conflict: Civilian Defense Militias and the Logic of Violence in Intrastate Conflict," *International Studies Perspectives*, 60, no. 3 (2016), 499–510; Jason Lyall, "Are Coethnics More Effective Counterinsurgents? Evidence from the Second Chechen War," *American Political Science Review*, 104, no. 1 (2010); Mario Fumerton, "Rondas Campesinas in the Peruvian Civil War: Peasant Self-Defence Organisations in Ayacucho," *Bulletin of Latin American Research*, 20, no. 4 (2001), 470–497.

23. The US cancelled $36,000 in military aid in 1994, then $0 for 1995 and 1996, then going back up to a minimal $205,000 in 1997: US Agency for International Development, "US Foreign Military Aid by Region and Selected Countries," last updated 2006. www.allcountries.org/ uscensus/1320_u_s_foreign_military_aid_by.html

24. Tim Weiner, "Tale of Evasion of Ban on Aid for Guatemala," *New York Times*, March 30, 1995.

25. US Department of State, *Guatemala Human Rights Practices, 1995* (1996); see also year before: US Department of State, *Guatemala Human Rights Practices, 1994* (1995).

26. Simone Remijnse, "Remembering Civil Patrols in Joyabaj, Guatemala," *Bulletin of Latin American Research*, 20, no. 4 (2001), 454–469.

27. E. San Juan, Jr., *US Imperialism and Revolution in the Philippines* (Basingstoke: Palgrave Macmillan, 2007); Alfred McCoy, *Policing America's Empire: The United States, the Philippines, and the Rise of the Surveillance State* (Madison, WI: University of Wisconsin Press, 2009).

28. McCoy, *Policing America's Empire*, 433–443; Erineo Espino, *Counter-insurgency: The Role of Paramilitaries* (Monterey, CA: Storming Media, 2004); James D. Ross, "Militia Abuses in the Philippines," *Third World Legal Studies*, 9, no. 1/7 (1990); David Kowalewski, "Counterinsurgent Paramilitarism: A Philippine Case Study," *Journal of Peace Research*, 29, no. 1 (February, 1992), 71–84.

29. Michael McClintock, *Instruments of Statecraft: U.S. Guerrilla Warfare, Counterinsurgency, and Counterterrorism, 1940–1990* (New York: Pantheon

Books, 1992); Ramsey Clark, *Right-Wing Vigilantes and US. Involvement: Report of a U.S.-Philippine Fact-Finding Mission to the Philippines* (Manila: Philippines Alliance of Human Rights Advocates, [1987]).

30. McCoy, *Policing America's Empire*, 436–441; Seth Mydans, "Right-Wing Vigilantes Spreading in Philippines," *New York Times*, April 4, 1987.

31. Amnesty International, *Amnesty International Report 1994—Philippines* (Amnesty International, [1994]).

32. Human Rights Watch, *They Own the People: The Ampatuans, State-Backed Militias, and Killings in the South Philippines* (New York: Human Rights Watch, 2010).

33. US Department of State. *Philippines Report on Human Rights Practices for 1996* (Washington, DC: US Department of State, 1997); Human Rights Solidarity, *Philippines: The Ramos Presidency and Human Rights—the Human Rights Record of the Ramos Administration from July 1992 to June 1997*, [1998]).

34. US Department of State. *The Philippines Country Report on Human Rights Practices for 1997* (Washington, DC: US Department of State, 1998).

35. Human Rights Watch, *They Own the People*, 22.

36. Katherine Hernandez, *Pirates in the Sea: Private Military and Security Company Activities in Southeast Asia and the Philippines Case* (Santiago, Chile: Global Consortium on Security Transformation, 2010); David Pugliese, "World: Soldiers of Fortune," *CorpWatch*, November 12, 2005; James Mann, "'Privatizing' of U.S. Bases in Philippines being Studied: Military: Turning Work Over to Contractors might Save Money. But Would it be Feasible for the Pentagon?" *LA Times*, April 13, 1990.

37. DynCorpInternational,"ABriefHistoryofDynCorpInternational,"*DynCorp International Website*, 2012. www.dyn-intl.com/about-us/history.aspx

38. Laura Peterson, *Making a Killing: Privatizing Combat, the New World Order* (Washington, DC: Center for Public Integrity, 2002).

39. Jorge Dominguez and Rafael Fernandez de Castro, *The United States and Mexico: Between Partnership and Conflict* (New York: Routledge, 2009, second ed.), 39–52; Graham H. Turbiville, *U.S. Military Engagement with Mexico: Uneasy Past and Challenging Future*, JSOU Report 10-2 (Hurlburt Field, FL: Joint Special Operations University, 2010), 14–17.

40. See, for example, Tom Hayden, ed., *The Zapatista Reader* (New York: Nation Books, 2002).

41. Julie Mazzei, *Death Squads Or Self-Defense Forces?: How Paramilitary Groups Emerge and Threaten Democracy in Latin America* (Chapel Hill, NC: University of North Carolina Press, 2009).

42. Ibid., 59.

43. Quoted in Mazzei, *Death Squads Or Self-Defense Forces?*, 58.

44. Kate Doyle, ed. *Breaking the Silence: The Mexican Army and the 1997 Acteal Massacre*, National Security Archive Electronic Briefing Book no. 283, National Security Archive, (2009), Document 2.

45. Diego Cevallos, "Report Links Paramilitaries with Ruling Party," *Inter Press Service*, April 30, 1999; Diego Cevallos, "Paramilitaries Attack Rebel Sympathizers," *Inter Press Service*, August 5, 2000; Diego Cevallos,

"Shadow of Paramilitaries Still Hangs Over Chiapas," *Inter Press Service*, September 30, 2001.

46. Mazzei, *Death Squads Or Self-Defense Forces?*, 47.

47. Marc Pugliese, "Acteal Massacre 1997," *Weiss Serota Helfman Pastoriza Cole & Boniske Website*. http://acteal97.com/?page_id=65 (accessed August 7, 2017); Andrew Kennis, "Ten Years Later, It's Time to Recognize the US Government's Responsibility for Acteal," *Narco News Bulletin*, December 30, 2007.

48. Heidi Moksnes, "Factionalism and Counterinsurgency in Chiapas: Contextualising the Acteal Massacre," *European Review of Latin American and Caribbean Studies*, 76 (April, 2004), 109–117.

49. Kristin Norget, "Caught in the Crossfire: Militarization, Paramilitarization, and State Violence in Oaxaca, Mexico," in Cecilia Menjivar and Nestor Rodriguez, eds., *When States Kill: Latin America, the US, and Technologies of Terror* (Austin, TX: University of Texas Press, 2005), 115–142; Peace Brigades International, *Human Rights Defenders in the State of Guerrero* (Peace Brigades International, 2007).

50. P. W. Singer, *Corporate Warriors* (Ithaca, NY: Cornell University Press, 2003), 40–48; Tony Geraghty, *Soldiers of Fortune: A History of the Mercenary in Modern Warfare* (New York: Pegasus Books, 2009); Peter W. Singer, "War, Profits, and the Vacuum of Law: Privatized Military Firms and International Law," *Columbia Journal of Transnational Law*, 42, no. 2 (2004), 521–549.

51. Christopher Kinsey, *Corporate Soldiers and International Security: The Rise of Private Military Companies* (London: Routledge, 2007), 50; Rita Abrahamsen and Michael C. Williams, "Security Beyond the State: Global Security Assemblages in International Politics," *International Political Sociology*, 3, no. 1 (2009); Rita Abrahamsen and Michael Williams, *Security Beyond the State: Private Security in International Politics* (Cambridge: Cambridge University Press, 2011); David Francis, "Mercenary Intervention in Sierra Leone: Providing National Security of International Exploitation?" *Third World Quarterly*, 20, no. 2 (1999), 319–323.

52. Colonel Bruce Grant, "US Military Expertise for Sale: Private Military Consultants as a Tool for Foreign Policy," *US Army War College* (1998), 6–7.

53. William Hartung, "Saudi Arabia: Mercenaries, Inc.," *The Progressive* (April 1, 1996), 2010.; Ken Silverstein, *Private Warriors* (London: Verso, 2000).

54. Ibid.

55. John E. Peck, "Remilitarizing Africa for Corporate Profit," *Z Magazine* (October 1, 2000); Singer, *Corporate Warriors*, 131.

56. Silverstein, *Private Warriors*.

57. Ibid., 182.

58. Deborah Kidwell, *Public War, Private Fight? The United States and Private Military Companies* (Fort Leavenworth, KS: Combat Studies Institute Press, 2005).

59. US Army, *Military Operations in Low Intensity Warfare: FM 100-20* (Washington, DC: Headquarters, Department of the Army, 1990).

60. Bruce E. Stanley, *Outsourcing Security: Private Military Contractors and US Foreign Policy* (Lincoln, NE: University of Nebraska Press, 2015); Allison Stanger, *One Nation Under Contract: The Outsourcing of American Power and the Future of Foreign Policy* (London: Yale University Press, 2009), 15.

61. Allison Stanger and Mark Williams, "Private Military Corporations: Benefits and Costs of Outsourcing Security," *Yale Journal of International Affairs* (Fall/Winter, 2006), 14; Kinsey, *Corporate Soldiers and International Security*; Singer, *Corporate Warriors*, 1; Deborah Avant, *The Market for Force: The Consequences of Privatising Security* (Cambridge: Cambridge University Press, 2005).

62. International Consortium of Investigative Journalists, *Making a Killing: The Business of War* (Washington, DC: Center for Public Integrity, 2003), 2.

63. Robert Mandel, *Armies without States* (London: Lynne Rienner, 2002), 8. These figures include a variety of services rendered including weapons construction, research and development, and other non-core activities. In addition, this number is what the US spent and does not include contracts hired by partner states, MNCs, and other actors.

64. Stanley, *Outsourcing Security*; Singer, *Corporate Warriors*, 80.

65. US-based PMCs required a license from the US government before they conduct operations, usually issued by the State Department: US Department of State, "Overview of U.S. Export Control System" (2011). www.state.gov/strategictrade/overview/; Foreign Commonwealth Office, "Private Military Companies: Options for Regulation," report presented to the House of Commons, London (February 2002); Ken Silverstein, "Privatising War," *The Nation*, 265 (1997), 120.

66. Silverstein, *Private Warriors*.

67. As quoted in Silverstein, "Privatising War."

68. It is important to note, for instance, that the majority of the emerging military contractors in the 1990s, such as MPRI or DynCorp, were run and staffed by former US Army personnel. MPRI once proudly claimed, for instance, that it had "more generals per square foot than the Pentagon"; MPRI, for example, is noted for its "loyalty to US foreign policy objectives," and the firm's headquarters are located a few miles from the Pentagon: Jeremy Scahill, *Blackwater: The Rise of the World's Most Powerful Mercenary Army* (New York: Nation Books, 2007); Nicolas Von Hoffman, "Contract Killers: How Privatizing the US Military Subverts Public Oversight" *Harper's Magazine* (June, 2004), 80.

69. Naomi Klein, *The Shock Doctrine* (London: Penguin Books, 2007), 308–325.

70. Michael Likosky, "The Privatisation of Violence," in S. Chesterman and A. Fisher, eds., *Private Security, Public Order* (Oxford: Oxford University Press, 2009), 18.

71. Kevin O'Brien, "PMCs, Myths, and Mercenaries," *RUSI Journal*, 145 (2000), 59–64; Kevin O'Brien, "Private Military Companies and African Security 1990–1998," in A. Musah and J. Fayemi, eds., *Mercenaries: An African Security Dilemma* (London: Pluto Press, 2000), 43–75; Thomas Adams, "The New Mercenaries and the Privatization of Conflict," *Parameters*, 29,

no. 2 (1999); Thomas Adams, "Private Military Companies: Mercenaries for the 21st Century," *Small Wars & Insurgencies*, 13, no. 2 (2002), 54–67.

72. Gerardo Renique, "'People's War,' 'Dirty War,': Cold War Legacy and the End of History in Postwar Peru," in Greg Grandin and Joseph Gilbert, eds., *A Century of Revolution* (Durham, NC: Duke University Press, 2010), 309–337.

73. Mario Fumerton and Simone Remijnse, "Civil Defense Forces: Peru's CAC and Guatemala's PAC in Comparative Perspective," in Kees Koonings and Dirk Kruijt, eds., *Armed Actors* (London: Zed Books, 2004), 52–72; Fumerton, "Rondas Campesinas in the Peruvian Civil War."

74. Fumerton and Remijnse, *Civil Defense Forces*; Cynthia McClintock, "The Decimation of Peru's Sendero Luminoso," in Cynthia J. Arnson, ed., *Comparative Peace Processes in Latin America* (Washington, DC: Woodrow Wilson Centre Press, 1999), 235–237.

75. Cynthia McClintock, "The United States and Peru in the 1990s," *George Washington University Working Papers* (2000); US Embassy, Lima, *1994 US State Department Report on Death Squad Operations in Peru*. Tamara Feinstein, ed., *Peru in the Eye of the Storm*, National Security Archive Electronic Briefing Book No. 64, National Security Archive, 2002).

76. Barbara Crossette, "US, Condemning Fujimori, Cuts Aid to Peru," *New York Times*, April 7, 1992; Anon., *Plans to Counteract Communism and Castro Penetration in Colombia and Other Parts of Latin America*. Memorandum of Conversation. Department of State, 1960. www.icdc. com/~paulwolf/colombia/gerberich15dec1960.htm (accessed July 20, 2012); Clifford Krauss, "US Military Team to Advise Peru in War Against Drugs and Rebels," *New York Times*, August 7, 1991; Johan Hayes, "U.S. Aid to Peru, all Programs, 1996–2001," *Just the Facts* (2014). http://justf.org/ Country%3Fcountry%3DPeru%26year1%3D1996%26year2%3D2014; Anthony Faiola, "US Allies in Drug War Disgrace," *Washington Post*, May 10, 2001.

77. Greg Guma, "The CIA, DynCorp, and the Shoot Down in Peru," *Zmag Online* (December, 2009). www.zcommunications.org/the-cia-dyncorp-and-the-shoot-down-in-peru-by-greg-guma; Charis Kamphuis, "Foreign Investment and the Privatization of Coercion: A Case Study of the Forza Security Company in Peru," *Brooklyn Journal of International Law*, 37, no. 1 (2011); Angela Paez, "Peru: UN Mission Probes Private Security Groups,." *Inter Press News Service*, February 7, 2007.

78. Silverstein, *Private Warriors*; Guma, "The CIA, DynCorp, and the Shoot Down in Peru."

79. Abrahamsen and Williams, *Security Beyond the State*, 126–135.

80. Afua Hirsch and John Vidal, "Shell Spending Millions of Dollars on Security in Nigeria, Leaked Data Shows," *Guardian*, August 19. 2012.

81. Silverstein, *Private Warriors*, 176; Adams, "The New Mercenaries and the Privatization of Conflict," 3.

82. John Levins, "The Kuwaiti Resistance," *Middle East Quarterly* (March, 1995); Frank Greve, "CIA, Army Said to be Supporting Kuwaiti Resistance," *Inquirer Washington Bureau*, August 31, 1990; Michael Wines,

"Confrontation in the Gulf: US is Said to Quietly Encourage a Kuwaiti Resistance Movement," *New York Times*, September 1, 1990.

83. Joel Brinkley, "The Reach of War: New Premier, Ex-CIA Aides Say Iraq Leader Helped Agency in 90s Attacks," *New York Times*, June 9, 2004.

84. Kenneth Katzman, *Iraq's Opposition Movements* (Washington, DC: Congressional Research Service, March 26, 1998); Major Paul Ott, *Unconventional Warfare in the Contemporary Operational Environment: Transforming Special Forces* (Fort Leavenworth, KS: School of Advanced Military Studies, 2001); Marianna Charountaki, *The Kurds and US Foreign Policy: International Relations in the Middle East since 1945* (Abingdon: Routledge, 2011).

85. For PMCs in support of US military in Operation Desert Storm, see Stanley, *Outsourcing Security*; David Shearer, 'Private Military Force and Challenges for the Future', *Cambridge Review of International Affairs*, no. 1 (1999), 80–94; Kidwell, *Public War, Private Fight?*.

86. See, for example, Doug Stokes, *America's Other War: Terrorizing Colombia* (New York: Zed Books, 2005), 122–124; Sam Raphael, "Paramilitarism and State Terror in Colombia," in Richard Jackson and S. Poynting, eds., *Contemporary State Terrorism: Theory and Practice* (London: Routledge, 2010).

87. Quoted in William Aviles, "The Political Economy of Low-Intensity Democracy: Colombia, Honduras, and Venezuela" in Ronald W. Cox, ed., *Corporate Power and Globalization in US Foreign Policy* (London: Routledge, 2012), 141.

88. Stokes, *America's Other War*.

89. See Human Rights Watch, *Colombia's Killer Networks: The Military-Paramilitary Partnership and the United States* (New York: Human Rights Watch, 1996). www.hrw.org/legacy/reports/1996/killertoc.htm; Human Rights Watch, "The History of the Military-Paramilitary Partnership," in *Colombia's Killer Networks*; Human Rights Watch, *The "Sixth Division": Military-Paramilitary Ties and U.S. Policy in Colombia* (New York: Human Rights Watch, 2001); Jasmin Hristov, *Blood and Capital: The Paramilitarization of Colombia* (Toronto: Between the Lines, 2009); Sam Raphael, "Paramilitarism and State Terror in Colombia," in Richard Jackson and S. Poynting, eds., *Contemporary State Terrorism: Theory and Practice* (London: Routledge, 2010); A. Weiss, "Colombia's Paramilitary: Profile of an Entrenched Terror Network," *Zmag Online* (April, 2001).

90. Hristov, *Blood and Capital*; Michael Evans, ed., "U.S. Intelligence Listed Colombian President Uribe Among 'Important Colombian Narco-Traffickers' in 1991," National Security Archive, August 2, 2004. http://nsarchive.gwu.edu/NSAEBB/NSAEBB131/

91. For the best analysis of these dynamics, see Hristov, *Blood and Capital*, 76–78; Stokes, *America's Other War*; Mazzei, *Death Squads Or Self-Defense Forces?*, 59.

92. Hristov, *Blood and Capital*, 12; Jasmin Hristov, "Uribe and the Paramilitarization of the Colombian State" *New Socialist*, 59 (2006), 14;

Jacobo Grajales, "The Rifle and the Title: Paramilitary Violence, Land Grab and Land Control in Colombia," *Journal of Peasant Studies*, 38, no. 4 (2011), 771-792.

93. Michael Evans, "Colombian Paramilitaries and the United States: 'Unraveling the Pepes Tangled Web': Documents Detail Narco-Paramilitary Connection to U.S.-Colombia Anti-Escobar Task Force," National Security Archive, Electronic Briefing Book No. 243, February 17, 2008. https://nsarchive2.gwu.edu//NSAEBB/NSAEBB243/index.htm

94. Best articulated by Hristov, *Blood and Capital*; and Stokes, *America's Other War*; Stokes, *Why the End of the Cold War Doesn't Matter*.

95. El Tiempo, "Asi Nacieron Las Convivir," *El Tiempo*, July 14, 1997. www.eltiempo.com/archivo/documento/MAM-605402; US Embassy, Bogota, "A Closer Look at Uribe's Auxiliary Forces," US Embassy in Bogota Cable to Secretary of State, Bogota, Colombia, September 11, 2002, in Michael Evans, ed., *Documents Implicate Colombian Government in Chiquita Terror Scandal*, National Security Archive Electronic Briefing Book No. 217, March 29, 2007. http://nsarchive.gwu.edu/NSAEBB/NSAEBB217/doc19.pdf; US Defense Intelligence Agency, "The Convivir in Antioquia—Becoming Institutionalized and Spreading Its Reach," Intelligence Information Report, April 7, 1997 in Evans, ed., *Documents Implicate Colombian Government in Chiquita Terror Scandal*. http://nsarchive.gwu.edu/NSAEBB/NSAEBB217/doc08.pdf

96. Stokes, *America's Other War*; Raphael, *Paramilitarism and State Terror in Colombia*; David Maher, "The Fatal Attraction of Civil War Economies: Foreign Direct Investment and Political Violence. A Case Study of Colombia," *International Studies Review*, 17, no. 2 (2015), 217–248; David Maher, "Rooted in Violence: Civil War, International Trade and the Expansion of Palm Oil in Colombia," *New Political Economy*, 20, no. 2 (2015), 299–330; David Maher and Andrew Thomson, "The Terror that Underpins the 'Peace': The Political Economy of Colombia's Paramilitary Demobilisation Process," *Critical Studies on Terrorism*, 4, no. 1 (2011), 95–113.

97. Michael Evans, ed., *The Chiquita Papers*, National Security Archive Electronic Briefing Book No. 340, April 7, 2011. http://nsarchive.gwu.edu/NSAEBB/NSAEBB340/; Evans, ed., *Documents Implicate Colombian Government in Chiquita Terror Scandal*. http://nsarchive.gwu.edu/NSAEBB/NSAEBB217/index.htm

98. Hristov, *Blood and Capital*, 77; Maher and Thomson, "The Terror that Underpins the 'Peace'"; Daniel Kovalik, "War and Human Rights Abuses: Colombia & the Corporate Support for Anti-Union Suppression," *Seattle Journal for Social Justice*, 2, no. 2 (2004).

99. Maher, *Rooted in Violence*; Maher, *The Fatal Attraction of Civil War Economies*; Maher and Thomson, "The Terror that Underpins the 'Peace.'"

100. As cited in Stokes, *Why the End of the Cold War Doesn't Matter*, 584.

101. However, recent media reports claim to have uncovered documents detailing CIA support for paramilitaries in Colombia in the 1990s leading up to the formation of the AUC: Brandon Barrett, "Paramilitary Emails

Allege CIA Worked with AUC," *Colombia Reports*, June 6, 2012; The CIA worked with *Los Pepes* the leader of which, Carlos Castano, later became the head of the AUC: Michael Evans, "Colombian Paramilitaries and the United States: 'Unraveling the Pepes' Tangled Web.'"

102. Maher and Thomson, "The Terror that Underpins the 'Peace.'"

103. Oliver Villar and Drew Cottle, *Cocaine, Death Squads, and the War on Terror: U.S. Imperialism and Class Struggle in Colombia* (New York: Monthly Review Press, 2011), 125.

104. See Stokes, *America's Other War*; Hristov, *Blood and Capital*.

105. See Hernando Calvo, "Colombia's Privatized Conflict," *Znet Online*, December 30, 2004.

106. Singer, *Corporate Warriors*, 207.

107. Government of Colombia, "Superintendencia De Vigilancia y Seguridad Privada," Government of Colombia. www.supervigilancia.gov.co/? (accessed April 14, 2013); Alvear Restrepo, "Private Security Transnational Enterprises in Colombia: Case Study Plan Colombia," *Corporación Colectivo De Abogados* (February, 2008).

108. Center for International Policy, "Report: Half of US Military Aid Goes through Private Contractors," *Center for International Policy Report Online*. www.cipcol.org/?p=416

109. See Irene Cabrera and Antoine Perret, "Colombia: Regulating Private Military and Security Companies in a 'Territorial State,'" *PIRV-WAR Report*, 19, no. 9 (November 15, 2009), 13.

110. Restrepo, "Private Security Transnational Enterprises in Colombia," 4; Singer, *Corporate Warriors*, 133.

111. Stanger and Williams, "Private Military Corporations: Benefits and Costs of Outsourcing Security", 10.

112. Senator Leahy (D-VT) as quoted in Paul de la Garza and David Adams. "Military Aid … from the Private Sector," *St. Petersburg Times*, December 3, 2000. For example, there are no legal mechanisms to guarantee that the content of military training provided by contractors conform to US human rights policies. Indeed, Patrick Leahy, the author of the Leahy Laws (the most stringent human rights legislation imposed on training of foreign military training), commented that "we have no way of knowing if the contractors are training these Colombian soldiers in ways that are fully consistent with U.S. policy, laws and procedures." Contractors hired by the CIA fall outside of all contract licensing controls: Stanger and Williams, "Private Military Corporations: Benefits and Costs of Outsourcing Security," 11.

113. For troop cap circumvention, see Kristen McCallion, "War for Sale! Battlefield Contractors in Latin America & the 'Corporatization' of America's War on Drugs," *U. Miami Inter-Am. L. Rev.*, 36 (Spring, 2005), 317–353; see also Grant, "US Military Expertise for Sale."

114. Stephan Fidler and Thomas Catn, "Colombia: Private Companies on the Frontline," *Financial Times*, August 12, 2003.

115. Myles Frechette, quoted in Jeremy Bigwood, "DynCorp in Colombia: Outsourcing the Drug War," *Corpwatch*, May 23, 2001.

116. Senator Leahy (D-VT), quoted in de la Garza and Adams, "Military Aid … from the Private Sector." This is different from a conventional understanding of "Plausible Denial," see McCallion, "War for Sale!"; Ruth Jamieson and Kieran McEvoy, "State Crime by Proxy and Judicial Othering," *British Journal of Criminology*, 45 (2005), 504–552; Stokes, *America's Other War*, 99.

117. Bigwood, "DynCorp in Colombia."

118. See, for instance, the graph in Restrepo, "Private Security Transnational Enterprises in Colombia," 8.

119. Ibid.

120. Singer, *Corporate Warriors*, 133.

121. Juan Tamayo, "Colombia: Private Firms Take on US Military Role in Drug War," *Miami Herald*, May 22, 2001; see also Bigwood, "DynCorp in Colombia."

122. See Calvo, *Colombia's Privatized Conflict*.

123. As quoted in Bigwood, "DynCorp in Colombia"; also see Calvo, *Colombia's Privatized Conflict*; Stanger and Williams, "Private Military Corporations: Benefits and Costs of Outsourcing Security", 9; Ross Eventon and Dave Bewley-Taylor, *Above the Law, Under the Radar: A History of Private Contractors and Aerial Fumigation in Colombia*, Global Drug Policy Observatory, Policy Report 4, February 2016. www.swansea.ac.uk/media/Privatisation_final.pdf

124. Restrepo, "Private Security Transnational Enterprises in Colombia."

125. Singer, *Corporate Warriors*, 208. For another example, see Tom Burghardt, "Did US Mercenaries Bomb the FARC Encampment in Ecuador?" *Global Research*, March 23, 2008; Fidler and Catn, "Colombia: Private Companies on the Frontline."

126. Bigwood, "DynCorp in Colombia."

127. See Steve Salisbury, "Pray and Spray: Colombia's Coke Bustin' Broncos," *Soldier of Fortune*, 23 (1998); Ignacio Gomez, "US Mercenaries in Colombia," *Colombia Journal*, July 16, 2000.

128. Juan Forero, "Private US Operatives on Risky Missions in Colombia," *New York Times*, February 14, 2004.

129. Northrop Grumman, "Northrop Grumman Statement to News Media regarding the Release of Ours Employees in Colombia," *Northrop Grumman*, 2010. www.irconnect.com/noc/press/pages/news_releases.html?d=145805

130. For more information on specific contracts and numbers, see Kristen McCallion, "War for Sale!"; Fidler and Catn, "Colombia: Private Companies on the Frontline"; Douglas Porch and Christopher Muller, "Imperial Grunts Revisited: The US Advisory Mission in Colombia," in Donald Stoker, ed., *Military Advising and Assistance: From Mercenaries to Privatization, 1815–2007* (New York: Routledge, 2010); Cabrera and Perret, "Colombia: Regulating Private Military and Security Companies in a 'Territorial State.'"

131. Mandel, *Armies without States*, 20; Singer, *Corporate Warriors*, 81.

132. Restrepo, "Private Security Transnational Enterprises in Colombia"; see also Government of Colombia, "Superintendencia De Vigilancia y Seguridad Privada."

133. Christian T. Miller, "A Colombian Town Caught in a Cross-Fire; the Bombing of Santo Domingo Shows How Messy U.S. Involvement in the Latin American Drug War Can Be," *LA Times*, March 17, 2002; Alan Axelrod, *Mercenaries: A Guide to Private Armies and Private Military Companies* (London: CQ Press, 2014); Tamayo, *Colombia: Private Firms Take on US Military Role in Drug War*.

134. Michael Gillard, Ignacio Gomez, and Melissa Jones, "BP Hands Tarred in Pipeline Dirty War," *Guardian*, October 17, 1998. Also see US Department of State, "Airscan International Inc.," US Department of State. www.state.gov/m/a/ips/c42178.htm (accessed July 6, 2012).

135. Matthew Reynolds, "Occidental Accused of Funding War Crimes," *Courthouse News Service*, November 1, 2011; Christian T. Miller, "Blood Spills to Keep Oil Wealth Flowing," *LA Times*, September 15, 2002.

136. Bigwood, "DynCorp in Colombia."

137. Calvo, *Colombia's Privatized Conflict*.

138. Brandon Barrett, "Israeli Mercenary Yair Klein Trained Paramilitary with the Approval of the Colombian Authorities," *Colombia Reports*, March 26, 2012.

139. Richard Norton-Taylor, "Drug Barons Army Trained by Britons," *Guardian*, December 6, 1990.

6. The War on Terror, Irregular Warfare, and the Global Projection of Force

1. Greg Grandin, *Empire's Workshop: Latin America, the United States, and the Rise of the New Imperialism* (New York: Owl Books, 2006).

2. Martin Shaw, *The New Western Way of War: Risk-Transfer War and its Crisis in Iraq.* (Cambridge: Polity Press, 2005). Also see Jeremy Kuzaramov, "Distancing Acts: Private Mercenaries and the War on Terror in American Foreign Policy," *Asia Pacific Journal*, 12/52, no. 1 (2014).

3. Charles Maier, "An American Empire," *Harvard Magazine*, 105, no 2 (2002); Michael Cox, "Empire, Imperialism, and the Bush Doctrine," *Review of International Studies*, 30, no. 4 (2004), 585–608; Michael Cox, "The Empire's Back in Town: Or America's Imperial Temptation Again," *Millennium*, 32, no. 1 (2003); John Bellamy Foster, "The Rediscovery of Imperialism," *Monthly Review* (November 2002). www.globalpolicy.org/component/content/article/154/25600.html

4. Gordon Lafer, "Neo-Liberalism by Other Means: The 'War on Terror' at Home and Abroad," in Joseph Preschek, ed., *The Politics of Empire: War, Terror, and Hegemony* (Abingdon: Routledge, 2006).

5. Richard Saull, *The Cold War and After: Capitalism, Revolution and Superpower Politics* (London: Pluto Press, 2007), 191.

6. David Kilcullen, "Countering Global Insurgency," *Journal of Strategic Studies*, 28, no. 4 (August, 2005), 597–617; Maria Ryan, "'Full Spectrum Dominance': Donald Rumsfeld, the Department of Defense, and US Irregular Warfare Strategy, 2001–2008," *Small Wars & Insurgencies*, 25, no. 1 (2014), 41–68.

7. The then Secretary of Defense Rumsfeld sent President Bush a series of memos even questioning the label the "War on Terror" due to the political and ideological nature of the threats posed to US interests: Donald Rumsfeld, "What Are We Fighting? Is It a Global War on Terror?," memo from Rumsfeld to George W. Bush, (June 18, 2004). library.rumsfeld. com/.../To%20President%20George%20W.%20Bush%20re%20Glob...

8. US Department of Defense, *2008 National Defense Strategy* (Washington, DC: Department of Defense, June, 2008). www.defense.gov/Portals/1/ Documents/pubs/2008NationalDefenseStrategy.pdf

9. See US Department of Defense, *The National Military Strategy of the United States of America* (Washington, DC: Department of Defense, 2011), 1–3; US Office of the President, *National Security Strategy* (Washington, DC: The White House, 2010); US Office of the President, *The National Security Strategy of the United States of America* (Washington, DC: The White House, 2002).

10. Michael Schwartz, *War Without End: The Iraq War in Context* (Chicago, IL: Haymarket Books, 2008); Doug Stokes and Sam Raphael, *Global Energy Security and American Hegemony* (Baltimore, MD: Johns Hopkins University Press, 2010), 96–97; William I. Robinson, *What to Expect from US 'Democracy Promotion' in Iraq* (Focus on the Global South, 2004). www.globalpolicy.org

11. Christopher Doran, *Making the World Safe For Capitalism: How Iraq Threatened the US Economic Empire and Had to Be Destroyed* (London: Pluto Press, 2012).

12. Max Fuller, "For Iraq, "The Salvador Option" Becomes a Reality," *Global Research Website*, June 2, 2005. http://globalresearch.ca/articles/FUL506A. html

13. Eric Herring and Glen Rangwala, *Iraq in Fragments: The Occupation and its Legacy* (London: C. Hurst and Co., 2006), 222–250.

14. Department of Defense, *2006 Quadrennial Defense Review Report*, 1.

15. Department of Defense, *Directive 3000.07* (2014); Department of Defense, *Directive 3000.07* (2008).

16. See, for instance, USJFCOM, *Irregular Warfare Special Study*, II-5. See also the Project for the New American Century, *Rebuilding America's Defenses: Strategy, Forces and Resources for a New Century* (Washington, DC, 2000); US Department of Defense, *Irregular Warfare: Directive 3000.07*, 2; US Army, *Army Special Operations Forces: Unconventional Warfare: FM 3-05.130*, 1-5, 3-19.

17. US Department of Defense, *Irregular Warfare Joint Operating Concept*, 9.

18. US Department of Defense, *Irregular Warfare: Directive 3000.07*, 2.

19. US Department of the Army, *Army Special Operations Forces: Unconventional Warfare: FM 3-05.130*, 1-3.

20. US Department of Defense, *Irregular Warfare Joint Operating Concept*, 32.

21. US Army, *Army Special Operations Forces: Unconventional Warfare: FM 3-05.130*, 2-11.

22 See, for instance, US Army, *Army Special Operations Forces: Unconventional Warfare: FM 3-05.130*, 3-11; USJFCOM, *Irregular Warfare Special Study*; Thomas H. Henriksen, "Afghanistan, Counterinsurgency and the Indirect Approach Report 10-3" (Hurlburt Field, FL: Joint Special Operations University Press, 2010).

23. US Department of Defense, *Irregular Warfare: Directive 3000.07*, 2; see also Richard C. Gross, *Different Worlds: Unacknowledged Special Operations and Covert Action*, 14. https://fas.org/man/eprint/gross.pdf (accessed March 14, 2018).

24. US Army, *Special Forces Unconventional Warfare Operations FM 3-05.201*; (Washington, DC: Department of the Army, 2003); US Army, *Army Special Operations Forces: Unconventional Warfare: FM 3-05.130* (Washington, DC: Department of the Army, September 2008); US Army, *Unconventional Warfare Mission Planning Guide for the Special Forces Operational Detachment—Alpha Level (TC 18-01.1)* (Washington, DC: Department of the Army, October, 2016); US Army, *Unconventional Warfare Pocket Guide* (Fort Bragg, NC: US Army Special Operations Command, Deputy Chief of Staff G3, Sensitive Activities Division G3X, AOOP-SA, 2016).

25. US Department of Defense, *Irregular Warfare: Directive 3000.07*, 11.

26. US Army, *Foreign Internal Defense Tactics, Techniques, and Procedures for Special Forces: FM 31-20-3*, 1-24.

27. US Army, *Tactics in Counterinsurgency: FM 3-24.2* (Washington, DC: Department of the Army, 2009), 3-22; US Army, *Foreign Internal Defense ATP 3-05.02* (Washington, DC: Department of the Army, August 2015); Armed Forces of the United States, *Joint Tactics, Techniques and Procedures for Foreign Internal Defense: Joint Publication 3-07.1*, glossary, 8; US Army, *Stability Operations: FM 3-07* (Washington, DC: Department of the Army, 2008); see also US Army, *Threat Force Paramilitary and Nonmilitary Organizations and Tactics: TC 31-93.3*, 1-38.

28. US Army, *Tactics in Counterinsurgency: FM 3-24.2* (Washington, DC: Department of the Army, 2009), 3-22.

29. Armed Forces of the United States, *Joint Tactics, Techniques and Procedures for Foreign Internal Defense: Joint Publication 3-07.1*, glossary, 8; US Army, *Tactics in Counterinsurgency: FM 3-24.2* (Washington, DC: Department of the Army, 2009), 3-22.

30. US Army, *Field Manual-Interim 3-07.22: Counterinsurgency Operations* (Washington, DC: Department of the Army, October 2004).

31. US Army, *Foreign Internal Defense Tactics, Techniques, and Procedures for Special Forces: FM 31-20-3*, Section 1-10 and pp. 1-19; US Army, *Army Special Operations Forces: FM 3-05* (Washington, DC: Department of the Army, 2006); US Army, *Special Forces Foreign Internal Defense Operations: FM 3-05.202*.

32. US Army, *Tactics in Counterinsurgency: FM 3-24.2* (Washington, DC: Department of the Army, 2009), 3-22.

33. This manual was reproduced as US Army, *Foreign Internal Defense Tactics, Techniques, and Procedures for Special Forces: FM 31-20-3* (Washington, DC: Department of the Army, 2004).

34. Quoted in Seth G. Jones, *Counterinsurgency in Afghanistan* (Santa Monica, CA: RAND Publications, 2008), 77.

35. US Army, *Tactics in Counterinsurgency: FM 3-24.2*, 8-12.

36. Lora Lumpe, "US Foreign Military Training: Global Reach, Global Power, and Oversight Issues," *Foreign Policy in Focus* (May, 2002); James S. Corum, "Training Indigenous Forces in Counterinsurgency: A Tale of Two Insurgencies," *Strategic Studies Institute* (March, 2006); Lt. Col. James Campbell, *Making Riflemen from Mud: Restoring the Army's Culture of Irregular Warfare* (Carlisle, PA: US Army War College, October 2007).

37. Richard A. Best, Jr. and Andrew Feickert, *Special Operations Forces (SOF) and CIA Paramilitary Operations: Issues for Congress* (Washington, DC: Congressional Research Service, 2009); see also Subcommittee on "Emerging Threats and Capabilities. The Future of U.S. Special Operations Forces: Ten Years after 9/11 and Twenty Five Years After the Goldwater-Nichols Hearing at House of Representatives," Washington, DC, September 22, 2011. https://fas.org/irp/congress/2011_hr/sof-future.pdf; Andrew Feikert, *U.S. Special Operations Forces (SOF): Background and Issues for Congress*, Congressional Research Service report, January 6, 2017. https://fas.org/sgp/crs/natsec/RS21048.pdf

38. Marcus Weisberger, "Peeling the Onion Back on the Pentagon's Special Operations Budget," *Defense One*, January 27, 2015; US Special Operations Command, *Operation and Maintenance, Defense-Wide Fiscal Year (FY) 2016 Budget Estimate*. http://comptroller.defense.gov/Portals/45/Documents/defbudget/fy2016/budget_justification/pdfs/01_Operation_and_Maintenance/O_M_VOL_1_PART_1/USSOCOM_PB16.pdf; USSOCOM, *FY2013 Budget Highlights United States Special Operations Command*; see also Feickert, *U.S. Special Operations Forces (SOF)*.

39. Nick Turse, "The U.S. Is Waging a Massive Shadow War in Africa, Exclusive Documents Reveal," *VICE News*, May 18, 2017.; Nick Turse, "American Special Operations Forces Are Deployed to 70 Percent of the World's Countries" *The Nation*, January 5, 2017.

40. US Congress, *Section 1208 Ronald W. Reagan National Defense Authorization Act for Fiscal Year 2005*, Public Law 108–375, 2004. www.gpo.gov/fdsys/pkg/PLAW-108publ375/pdf/PLAW-108publ375.pdf; see also Kibbe, *The Rise of the Shadow Warriors*, 103.

41. Colby Goodman, Christina Arabia, Robert Watson, Taner Bertuna, and Andrew Smith, *US Foreign Military Training Reached Record Highs in 2015*. Center for International Policy. Security Assistance Monitor Trend Report (May 2017). www.ciponline.org/images/uploads/actions/FMTR2015_SAM_052017_1.pdf (accessed November 8, 2017); Douglas Gillison, Nick Turse, and Moiz Syed, "The Network: Leaked Data Reveals How the U.S. Trains Vast Numbers of Foreign Soldiers and Police With Little Oversight," *The Intercept* (blog), July 13, 2016. https://theintercept.com/2016/07/13/training/; Nick Turse, "American Special Operations Forces Are Deployed

to 70 Percent of the World's Countries," *The Nation*, January 5, 2017; Federation of American Scientists, "US International Security Assistance Education and Training," Federation of American Scientists, n.d.; US Department of State and US Department of Defense, *Foreign Military Training Fiscal Years 2015 and 2016 Joint Report to Congress.* Washington, DC. www.state.gov/documents/organization/265162.pdf; Nick Turse, "Generals and Cops Trained by the Pentagon Are Staging Coups All Over the World," *In These Times*, August 10, 2017; John Norris, "Is America Training Too Many Foreign Armies?" *Foreign Policy*, January 28, 2013. http://foreignpolicy.com/2013/01/28/is-america-training-too-many-foreign-armies/ (accessed November 8, 2017); Eric Schmitt, and Tim Arango, "Billions From U.S. Fail to Sustain Foreign Forces," *New York Times*, October 3, 2015.

42. Richard Davis, ed., *The US Army and Irregular Warfare 1775–2007* (Washington, DC: US Government Printing Office, 2007). For an overview and critique of this, see Vacca and Davidson, *The Regularity of Irregular Warfare*, 18–28; for iterations of this being a new paradigm, see Jeffrey B. White, "Some Thoughts on Irregular Warfare: A Different Kind of Threat," *CIA*, www.cia.gov/library/center-for-the-study-of-intelligence/csi-publications/csi-studies/studies/96unclass/iregular.htm (accessed February 16, 2011); USJFCOM, *Irregular Warfare Special Study*; Newton, *The Seeds of Surrogate Warfare*, 1–19; Peltier, *Surrogate Warfare*, 55–58.

43. Campbell, *Making Riflemen from Mud*, 5.

44. For further information on the wider adoption of counterinsurgency and unconventional warfare in conventional military structures, see David Ucko, *The New Counterinsurgency Era: Transforming the US Military for Modern Wars* (Washington, DC: Georgetown University Press, 2009); Robert M. Cassidy, *Counterinsurgency and the Global War on Terror* (New York: Greenwood Publishing, 2006).

45. Deborah Kidwell, "Public War, Private Fight? The United States and Private Military Companies," *Global War on Terrorism Occasional Paper 12.* (Fort Leavenworth, KS: Combat Studies Institute Press, 2005), 29; Moshe Schwartz and Joyprada Swain, *Department of Defense Contractors in Afghanistan and Iraq: Background and Analysis* (Washington, DC: Congressional Research Service, 2011), 9.

46. Moshe Schwartz, John F. Sargent, Gabriel Nelson, and Ceir Coral, "Defense Acquisitions: How and Where DOD Spends and Reports Its Contracting Dollars." *Congressional Research Service Report*, December 20, 2016. https://fas.org/sgp/crs/natsec/R44010.pdf

47. Deborah Avant, *The Market for Force: The Consequences of Privatising Security*, (Cambridge: Cambridge University Press, 2005), 122; Donald Stoker, ed., *Military Advising and Assistance: From Mercenaries to Privatization, 1815–2007* (New York: Routledge, 2008).

48. Adam Moore, "US Military Logistics Outsourcing and the Everywhere of War," *Territory, Politics, Governance*, 5 no. 1 (2017), 5–27; Sean McFate, *The Modern Mercenary: Private Armies and What They Mean for World Order* (Oxford: Oxford University Press, 2014), 44; Bruce E. Stanley,

Outsourcing Security: Private Military Contractors and US Foreign Policy (Lincoln, NE: University of Nebraska Press, 2015), 9.

49. For reports on specific proportions of contract service types hired by DoD, see Kidwell, "Public War, Private Fight?," 30, and Schwartz and Swain, *Department of Defense Contractors in Afghanistan and Iraq*, 16.

50. Lora Lumpe, "US Foreign Military Training: Global Reach, Global Power, and Oversight Issues," *Foreign Policy in Focus* (May, 2002), 12.; P. W. Singer, "War, Profits, and the Vacuum of Law: Privatized Military Firms and International Law," *Columbia Journal of Transnational Law*, 42, no. 2 (2004): 521–549; Colonel Bruce Grant, "US Military Expertise for Sale: Private Military Consultants as a Tool for Foreign Policy" (Washington, DC: US Army War College, 1998).

51. Lumpe, *US Foreign Military Training*; Stoker, *Military Advising and Assistance*.

52. Michael Guidry and Guy J. Wills, "Future UAV Pilots: Are Contractors the Solution?" *Air Force Journal of Logistics*, 28, no. 4 (2004).

53. Kyle Ballard, "The Privatization of Military Affairs: A Historical Look into the Evolution of the Private Military Industry," in Thomas Jäger and Gerhard Kümmel, eds., *Private Military and Security Companies* (Wiesbaden: VS Verlag, 2007), 51; P. W. Singer, *Corporate Warriors* (Ithaca, NY: Cornell University Press, 2003), 16.

54. David Cloud, "Civilian Contractors Playing Key Roles in US Drone Operations," *LA Times*, December 29, 2011.

55. Antonio Taguba, *The "Taguba Report" on Treatment of Abu Ghraib Prisoners in Iraq: Article 15-6*. https://fas.org/irp/agency/dod/taguba.pdf; Simon Chesterman, "We can't Spy … if we can't Buy!': The Privatization of Intelligence and the Limits of Outsourcing 'Inherently Governmental Functions," *European Journal of International Law*, 19, no. 5 (2008), 1055; Josh Meyer, "CIA Contractors Will be a Focus of Interrogation Investigation," *LA Times*, August 27, 2009; Renae Merle and Ellen McCarthy, "6 Employees from CACI International, Titan Referred for Prosecution," *Washington Post*, August 26, 2004.

56. See Avant, *The Market for Force*, 122; for official figures, see Commission on Wartime Contracting, *At What Risk? Correcting Over-Reliance on Contractors in Contingency Operations* (Washington, DC: Commission on Wartime Contracting, 2011).

57. Isenberg, *Shadow Force*, 76.

58. Jim Vallette and Pratap Chatterjee, "Guarding the Oil Underworld in Iraq," *CorpWatch* 2003. www.corpwatch.org/article.php?id=8328

59. See Pratap Chatterjee, *Iraq, Inc.: A Profitable Occupation* (New York: Seven Stories Press, 2004), 119; Human Rights Watch, *Well Oiled Oil and Human Rights in Equatorial Guinea* (New York: Human Rights Watch, 2009), 83.

60. Avant, *The Market for Force*, 122.

61. Mark Mazzetti, "C.I.A. Sought Blackwater's Help to Kill Jihadists," *New York Times*, August 19, 2009. www.nytimes.com/2009/08/20/us/20intel. html (accessed March 15, 2012); Mark Mazzetti and Scott Shane, "C.I.A. had Plan to Assassinate Qaeda Leaders," *New York Times*, July 13, 2009.

62. Ian Cobain and Ben Quinn, "How US Firms Profited from Torture Flights," *Guardian*, August 31, 2011; Grey, *Ghost Plane: The Inside Story of the CIA's Secret Rendition Program*; Ruth Blakeley and Sam Raphael, "The Rendition Project." www.therenditionproject.org.uk/index.html (accessed November 5, 2017).

63. Bob Woodward, *Bush at War* (New York: Simon and Schuster, 2002), 141–146. For a personal account of these events, see Gary Bernstein and Ralph Pezzullo, *Jawbreaker: The Attack on Bin Laden and Al-Qaeda: A Personal Account by the CIA's Key Field Commander* (New York: Crown Publishing Group, 2006); Gary C. Schroen, *First In: An Insider's Account of how the CIA Spearheaded the War on Terror in Afghanistan* (New York: Presidio Press, 2005), 88.

64. Bob Woodward, "CIA Led Way with Cash Handouts," *Washington Post*, November 18, 2002.

65. Antonio Giustozzi, *Empires of Mud: Wars and Warlords in Afghanistan* (New York: Columbia University Press, 2009), 89.

66. Woodward, "CIA Led Way with Cash Handouts."

67. Henriksen, *Afghanistan, Counterinsurgency and the Indirect Approach*, 39.

68. See ibid.

69. Giustozzi, *Empires of Mud*; see also Jones, *Counterinsurgency in Afghanistan*, 33–35.

70. Giustozzi, *Empires of Mud*, 88–91; Peter Tomsen, *The Wars of Afghanistan* (Philadelphia, PA: Public Affairs, 2011), 643–648; Human Rights Watch, *Afghanistan: Return of the Warlords* (New York: Human Rights Watch, 2002).

71. See, for instance, Major Jim Gant, *One Tribe at a Time* (Los Angeles, CA: Nine Sisters Imports, Inc., 2009), 11; Small Wars Foundation, "Tribal Engagement Workshop," *Small Wars Journal*, Conference held at Gari Melcher's Home and Studio, Fredericksburg, VA, March 24–25 2010. http://smallwarsjournal.com/content/tribal-engagement-workshop; Seth G. Jones, *The Strategic Logic of Militia* (Santa Monica, CA: RAND, 2012); Major John D. Litchfield, "Unconventional Counterinsurgency: Leveraging Traditional Social Networks and Irregular Forces in Remote and Ungoverned Areas," *School of Advanced Military Studies* (2010); Burris, Bradford. "Applying Iraq to Afghanistan," *Small Wars Journal* (2010); Yochi Dreazen, "US to Fund Afghan Militias, Applying Iraq Tactic," *Wall Street Journal*, December 23, 2008; Jon Boone, "US Keeps Secret Anti-Taliban Militia on a Bright Leash," *Guardian*, March 8, 2010.

72. Spencer Ackerman, "A CIA COINdinista's Misgivings on Counterinsurgency in Afghanistan," *Washington Independent*, May 13, 2010.

73. Henriksen, *Afghanistan, Counterinsurgency and the Indirect Approach*, 76; Antonio Giustozzi, "The Privatizing of War and Security in Afghanistan: Future Or Dead End?" *Economics of Peace and Security Journal*, 2, no. 1 (2007), 30.

74. Greg Miller, "CIA Expanding Presence in Afghanistan," *LA Times*, September 20, 2009; Mark Sedra, "Small Arms and Security Sector Reform," in Michael Bhaktia and Mark Sedra, eds., *Afghanistan, Arms and*

Conflict: Armed Groups, Disarmament and Security in a Post-War Society (Abingdon: Routledge, 2008), 176.

75. UPI, "US Funds Sons of Iraq," *UPI*, October 29, 2009.

76. Seth Jones and Arturo Munoz, *Afghanistan's Local War: Building Local Defense Forces* (Santa Monica, CA: RAND Corporation, 2010); Jon Boone, "US Pours Millions into Anti-Taliban Militias in Afghanistan," *Guardian*, November 22, 2009.

77. Kenneth Katzman, *Afghanistan: Post-Taliban Governance, Security, and U.S. Policy* (Washington, DC: Congressional Research Service, 2012); Giustozzi, *Empires of Mud*, 92; Jones, *Counterinsurgency in Afghanistan*, 79–80; Human Rights Watch, *Just Don't Call it a Militia: Impunity, Militias, and the "Afghan Local Police"* (New York: Human Rights Watch, 2011), 15–18.

78. Jones and Munoz, *Afghanistan's Local War*; Katzman, *Afghanistan: Post-Taliban Governance, Security, and U.S. Policy.*

79. Center for Army Lessons Learned, *Southern Afghanistan COIN Operations no. 07-6* (Washington, DC: US Department of the Army, 2006), 29.

80. Human Rights Watch, *Just Don't Call it a Militia*, 18–19. Andrew Wilder, *Cops Or Robbers? The Struggle to Reform the Afghan National Police* (Kabul: Afghanistan Research and Evaluation Unit, 2007); Jones, *Counterinsurgency in Afghanistan*, 76.; Wilder, *Cops Or Robbers?* 13–14.

81. See Human Rights Watch, *Just Don't Call it a Militia*, 22–53.

82. President Karzai as cited in Human Rights Watch, *Just Don't Call it a Militia*, 53.

83. Katzman, *Afghanistan: Post-Taliban Governance, Security, and U.S. Policy*, 37.

84. Human Rights Watch, *Just Don't Call it a Militia*; Ernesto Londono, "U.S. Cites Local Afghan Police Abuses," *Washington Post*, December 16, 2011; Yaroslav Trofimov, "Afghan Militia Wins Uneasy Peace," *Wall Street Journal*, May 29, 2012; Dan De Luce, "Pentagon Defends Afghan Local Police Program," *AFP*, May 14, 2012.

85. See, for instance, Phillip Alston. "Special Rapporteur of the United Nations Human Rights Council on Extrajudicial, Summary or Arbitrary Executions," Press Statement (Kabul, May 15, 2008). www.extrajudicialexecutions.org/application/media/Statement,%2015%20May%202008,%20Kabul,%20Afghanistan%20%5BEnglish%5D.pdf; Jerome Starkey, "Afghan Death Squads Acting on Foreign Orders," *Independent*, May 16, 2008.

86. See, for instance, DynCorp International, "DynCorp International Awarded Mentoring and Training Contract in Afghanistan," www.dyn-intl.com; Christine Spolar, "DynCorp Wins Contract Dispute Over Afghan Police Training," *Huffington Post*, May 25, 2011.

87. Giustozzi, *The Privatizing of War and Security in Afghanistan*; Kinsey, *Corporate Soldiers and International Security*, 23.; White, Andrew. "Afghan Special Operations Command to Double in Size—SOF—Special Operations," *Shephard Media*, September 1, 2017. www.shephardmedia.com/news/special-operations/afghan-special-operations-command-double-size/ (accessed November 8, 2017).

88. Representative John Terney, *Warlord, Inc.: Extortion and Corruption Along the U.S. Supply Chain in Afghanistan* (Washington, DC: Committee on Oversight and Government Reform, US House of Representatives, 2010).

89. Committee on Armed Services, US Senate, *Inquiry into the Role and Oversight of Private Security Contractors in Afghanistan*. https://fas.org/irp/congress/2010_rpt/sasc-psc.pdf; McFate, Sean, *The Modern Mercenary: Private Armies and What They Mean for World Order* (Oxford: Oxford University Press, 2014), 156–157.

90. US Department of Defense, *Contractor Support of U.S. Operations in the USCENTCOM Area of Responsibility* (July 2016). www.acq.osd.mil/log/ps/.CENTCOM_reports.html/5A_July_2016_Final.pdf

91. Leo Shane III, "Report: Contractors Outnumber US Troops in Afghanistan 3-1," *Military Times*, August 17, 2016.

92. Katrina Manson, "Erik Prince Offers Private Military Force in Afghanistan," *Financial Times*, August 7, 2017.

93. Sune Engel Rasmussen, "UN Concerned By Controversial Plan to Revive Afghan Militias," *Guardian*, November 19, 2017.

94. Woodward, *Plan of Attack*, Chapter 10.

95. Dana Priest and Josh White, "Before the War, the CIA Reportedly Trained a Team of Iraqis to Aid US," *Washington Post*, August 3, 2005. www.informationclearinghouse.info/article9660.htm; Mike Tucker and Charles Faddis, *Operation Hotel California* (Guilford: The Globe Pequat Press, 2009), 33–38.

96. Woodward, *Plan of Attack*; Steven Hurst, *The United States and Iraq since 1979: Hegemony, Oil, and War* (Edinburgh: Edinburgh University Press, 2009), 129–131; Hirsh and Barry, "The Salvador Option."

97. Isenberg, *Shadow Force*, 19.

98. Robert Dreyfuss, "Phoenix Rising," *American Prospect*, January 1, 2004. http://prospect.org/article/phoenix-rising

99. Hirsh and Barry, "The Salvador Option."

100. See, for instance, Grandin, *Empire's Workshop*; Scahill, *Blackwater*, 349. See also Brussels Tribunal. "The Salvador Option Exposed." www.brussellstribunal.org/BritishBombers.htm; Peter Maass, "The Way of the Commandos," *New York Times Magazine*, May 1, 2005.

101. Scahill, *Blackwater*, 377.

102. Fuller, *For Iraq, "The Salvador Option" Becomes a Reality*; Scahill, *Blackwater*, 352.

103. BBC News, "Profile: Iraq's Wolf Brigade," *BBC News*, June 11, 2006; Max Fuller, "Crying Wolf: Media Disinformation and Death Squads in Occupied Iraq," *Global Research*, November 10, 2005; Gareth Porter, "US Military Still Runs with Dreaded Wolf Brigade," *IPS News*, January 2, 2006.

104. Schwartz, *War Without End: The Iraq War in Context*, 211; Fuller, *For Iraq, "The Salvador Option" Becomes a Reality*; for a broader study on the intersection on the perpetuation of state terror and neoliberalism promotion in the South, Ruth Blakeley, *State Terrorism and Neoliberalism: The North in the South* (London: Routledge, 2011).

105. See, for instance Fuller, *For Iraq, "The Salvador Option" Becomes a Reality*; Fuller, "Crying Wolf: Media Disinformation and Death Squads in Occupied Iraq"; Monitoring of Human Rights in Iraq (MHRI), "Death Squads in Iraq: Evolution, Objectives, Results," (Baghdad: MHRI, December 2006). www.brusselstribunal.org/pdf/deathSquadsMHRI.PDF; Shane Bauer, "Iraq's New Death Squad," *The Nation*, 6 (June, 2009).

106. Fuller, *For Iraq, "The Salvador Option" Becomes a Reality.*

107. Stokes and Raphael, *Global Energy Security and American Hegemony*, 103.

108. Amnesty International, *Iraq: Civilians Under Fire* (London: Amnesty International, 2010); Steven Harris, "Who's Behind the Active Death Squads Running Iraq?" *Al Jazeera*, July 20, 2006; J. Cogan, "Journalist Killed After Investigating US-Backed Death Squads in Iraq," *World Socialist Web Site*, July 1, 2005. www.wsws.org/articles/2005/jul2005/iraq-j01.shtml

109. Amy Goodman, "Exclusive: Former UN Human Rights Chief in Iraq Says US Violating Geneva Conventions, Jailing Innocent Detainees," *Democracy Now*, February 28, 2006.

110. Harris, "Who's Behind the Active Death Squads Running Iraq?"

111. Human Rights Watch. *Iraq: End Interior Ministry Death Squads* (New York: Human Rights Watch, 2006).

112. Ibid.; Lionel Beehner, *Iraq's Militia Groups*, Council on Foreign Relations, October 26, 2005.

113. Seymour Hersh, "Moving Targets," *The New Yorker*, December 15, 2003.

114. Avant, *The Market for Force*, 122.

115. Northrop Grumman, "Northrop Grumman Statement to News Media regarding the Release of Our Employees in Colombia," 2010. www.irconnect.com/noc/press/pages/news_releases.html?d=145805

116. David Isenberg, "DynCorp in Iraq," *Huffington Post*, January 24, 2010; August Cole, "DynCorp's Iraq Contracts with U.S. are Scrutinized," *Wall Street Journal*, January 25, 2010.

117. Pratap Chatterjee, "The Boys from Baghdad: Iraqi Commandos Trained by US Contractor," *CorpWatch.Com*, September 20, 2007. www.corpwatch.org/article.php?id=14700

118. Defense Industry Daily "MARSS & More: Quasi-Civilian Spy Plane Services on the Front Lines," *Defense Industry Daily*, March 31, 2011.

119. US Department of Defense, "Contracts." www.defense.gov/contracts/contract.aspx?contractid=4501; Isenberg, *Shadow Force*, 74.

120. Isenberg, *Shadow Force*. Also see Chesterman, "'We can't Spy … if we can't Buy!'" 1055.

121. Scahill, *Blackwater*, 110; Jeremy Scahill, "Blackwater: CIA Assassins?" *The Nation*, August 20, 2009.

122. Sarah Meyer, "Iraq: Security Companies and Training Camps," *Global Research*, May 18, 2006. www.globalresearch.ca/index.php?context=va&aid=2461#; Mandel, *Armies Without States*, 20.

123. Vallette and Chatterjee, *Guarding the Oil Underworld in Iraq*.

124. For more on the FPS, see Global Security.org "Facility Protection Service." www.globalsecurity.org/intell/world/iraq/fps.htm (accessed July 7, 2012).

125. Stephen Armstrong, *War PLC: Rise of the New Corporate Mercenary* (Chatham: Faber and Faber, 2008), 39.

126. Steve Fainaru and Alec Klein, "In Iraq, A Private Realm of Intelligence Gathering," *Washington Post*, July 1, 2007.

127. Isenberg, *Shadow Force*; David M. Barnes, *The Ethics of Military Privatization: The US Armed Contractor Phenomenon* (London: Routledge, 2016).

128. Govinda Clayton and Andrew Thomson, "The Enemy of My Enemy is My Friend … the Dynamics of Self Defense Forces in Irregular War: The Case of the Sons of Iraq," *Studies in Conflict & Terrorism*, 37, no. 11 (2014), 920–935; William McCallister, "Sons of Iraq: A Study in Irregular Warfare," *Small Wars Journal* (September 8, 2008).

129. Schwartz, *War Without End*, 241; also see John A. McCary, "The Anbar Awakening: An Alliance of Incentives," *Washington Quarterly*, 32, no. 1 (January, 2009), 43–59.

130. Schwartz, *War Without End*, 246–247; Office of the Special Inspector General for Iraq Reconstruction, "Sons of Iraq Program: Results are Uncertain and Financial Controls are Weak," SIGIR, January 28, 2011.

131. Federico Manfredi, "Iraq's Crooked Politicians: Talking with Sheik Ali Hatem, Leader of the Sunni Awakening Councils," *Huffington Post*, August 16, 2010.

132. Greg Bruno, "The Role of the "Sons of Iraq" in Improving Security," *Washington Post*, April 28, 2008.

133. Ahram, *Proxy Warriors*, 90.

134. Shane Bauer, "Iraq's New Death Squad," *The Nation*, 6 (June, 2009); Human Rights Watch "Iraq: Pro-Government Militias' Trail of Death," July 31, 2014. www.hrw.org/news/2014/07/31/iraq-pro-government-militias-trail-death; Amnesty International "Absolute Impunity: Militia Rule in Iraq" (London: Amnesty International 2014). www.amnesty.org.uk/files/absolute_impunity_iraq_report.pdf

135. Molly Hennessy-Fiske and W. J. Hennigan, "The U.S. Is Helping Train Iraqi Militias Historically Tied to Iran," *Los Angeles Times*, December 14, 2016. www.latimes.com/world/middleeast/la-fg-iraq-shiite-militias-20161212-story.html; Ibrahim Al-Marashi, "The Future of Militias in Post-ISIL Iraq," *Al-Jazeera*, March 27, 2017. www.aljazeera.com/indepth/opinion/2017/03/future-militias-post-isil-iraq-170324090705631.html (accessed November 8, 2017); Jack Watling, "The Shia Militias of Iraq," *The Atlantic*, December 22, 2016. www.theatlantic.com/international/archive/2016/12/shia-militias-iraq-isis/510938/

136. Asharq Al-Awsat, "Middle-East Arab News Opinion," Asharq Al-Awsat English Archive, May 3, 2015. https://eng-archive.aawsat.com/hamza mustafa/news-middle-east/iraqi-parliament-rejects-us-bill-to-split-iraq; David Patel, "ISIS in Iraq: What We Got Wrong and Why 2015 is not 2007 Redux," *Middle East Brief*, no. 87, Brandeis University (January 2015). www.brandeis.edu/crown/publications/meb/MEB87.pdf

137. US Department of Defense "Contractor Support of U.S. Operations in the USCENTCOM Area of Responsibility," January 2016. www.acq.osd.

mil/log/PS/.CENTCOM_reports.html/5A_January_2016_Final.pdf; Strobel, Warren and Phil Stewart "As U.S. Troops Return to Iraq, More Private Contractors Follow," *Reuters*, December 24, 2014. www.reuters. com/article/us-usa-iraq-contractors/as-u-s-troops-return-to-iraq-more-private-contractors-follow-idUSKBN0K20AW20141224; Marcus Weisberger, "Back to Iraq: US Military Contractors Return In Droves," *Defense One*, February 2016. www.defenseone.com/threats/2016/02/back-iraq-us-military-contractors-return-droves/126095/ (accessed November 8, 2017); Kelley Beaucar Vlahos, "Rise of the American Mercenary," *The American Conservative*. www.theamericanconservative.com/articles/rise-of-the-american-mercenary/ (accessed November 8, 2017).

138. Jeremy Scahill, "Washington's War in Yemen Backfires," *The Nation*, February 15, 2012; Jeremy Sharp, *Yemen: Background and U.S. Relations* (Washington, DC: Congressional Research Service, February 11, 2015); Stephen Zunes, "How the US Contributed to Yemen's Crisis," *Foreign Policy in Focus*, April 20, 2015.

139. Sudarsan Raghavan, "In Yemen, Tribal Militias in a Fierce Battle with Al-Qaeda Wing," *Washington Post*, September 10, 2012; Casey Coombs, "Echos of Iraq: Yemen's War Against Al-Qaeda Takes a Familiar Turn," *Time Magazine*, August 10, 2012; Farea Al-Muslimi, "The Popular Committee Phenomenon in Yemen: Fueling War and Conflict," *Carnegie Middle East Center*, April 1, 2015. http://carnegie-mec.org/diwan/59560?lang=en; Sasha Gordon, "Tribal Militias in Yemen: Al Bayda and Shabwah," *American Enterprise Institute*, February 6, 2013; Muhammad Al-Abasi, "Bi'l-Watha'iq fasad wizarat aldifa al-yamaniya yaltahim mukhassasat al-lijan al-shabiya fi Abyan" ("Documentary Proof of Corruption in Yemen's Ministry of Defense Gobbling Up the Allocations for the Popular Committees in Abyan"), *Sabq News (Yemen)*, March 7, 2013.

140. Emily Hager and Mark Mazzetti, "Emirates Secretly Sends Colombian Mercenaries To Yemen Fight," *New York Times*, November 25, 2015; Mark Mazzetti and Emily Hager, "Secret Desert Force Set up by Blackwater's Founder," *New York Times*, May 14, 2011; Laura Carlsen, "Mercenaries in Yemen—The U.S. Connection," *Huffington Post*, December 3, 2015; Middle East Eye, "Yemeni mercenaries fight and die to protect Saudi border" *Middle East Eye*, August 4, 2017. www.middleeasteye.net/news/yemenis-fight-saudi-arabia-s-war-return-pennies-1458536939

141. Thomas Gibbons-Neff and Missy Ryan, "U.S. Special Operations Extends Yemen Mission Against Al-Qaeda," *Washington Post*, June 17, 2016; Carlo Muñoz, "U.S. Military in Yemen Supporting UAE, Saudi Arabia in Fight Against al Qaeda," *Washington Times*, May 6, 2016.

142. Juan Cole, "Trump's Generals Are Considering a Wider War in Yemen," *The Nation*, April 3, 2017; *American Herald Tribune* Staff, "US Hiring Mercenaries with $1,500 Daily Wages for War in Southern Yemen: Reports," *American Herald Tribune*, February 13, 2017.

143. Jane Perlez and Pir Zubair Shah, "Pakistan Uses Tribal Militia in Taliban War," *New York Times*, October 23, 2008; Karin Brulliard, "Pakistani Anti-Taliban Militias Offer Lessons for U.S. in Afghanistan," *Washington Post*,

December 7, 2010; Salman Yousafzi, "Death by Lakshar: The Forgotten Protectors of Adezai Village," *Dawn*, May 9, 2016; Yelena Biberman, "Reimagining Pakistan's Militia Policy," *Atlantic Council*, April 2015.

144. Julius Cavendish, "How the CIA Ran a Secret Army of 3,000 Assassins," *Independent*, September 22, 2010; Sudarsan Raghavan, "CIA Runs Shadow War With Afghan Militia Implicated in Civilian Killings," *Washington Post*, December 3, 2015; Dexter Filkins and Mark Mazzetti, "Contractors Tied to Effort to Track and Kill Militants," *New York Times*, 14 March, 2010; Bob Woodward, *Obama's Wars* (New York: Simon and Schuster, 2011); Jeremy Scahill, "The Secret U.S. War in Pakistan," *The Nation*, November 23, 2009; Mark Mazzetti, "CIA Sought Blackwater's Help to Kill Jihadists," *New York Times*, August 19, 2009. DynCorp was also given a contract to protect Pakistani diplomats, prompting concerns among local officials that it was being used as a cover for the development of a parallel intelligence structure: Jane Perlez, "U.S. Push to Expand in Pakistan Meets Resistance," *New York Times*, October 6, 2009; Jeremy Scahill, *Dirty Wars: The World is a Battlefield* (New York: The Nation Books, 2013), 415; Mark Mazzetti, "How a Single Spy Helped Turn Pakistan Against the U.S.," *New York Times*, April 9, 2013.

145. Rob Prince and Ibrahim Kazerooni, "Syria, the United States, and the El Salvador Option," *Foreign Policy in Focus*, June 20, 2012 (parts 1 and 2). http://fpif.org/syria_the_united_states_and_the_el_salvador_option_part_one/; Ariel Zirulnik, "Cables Reveal Covert US Support for Syria's Opposition," *Christian Science Monitor*, April 18, 2011; Michael Chussodovsky "'The Salvador Option For Syria': US-NATO Sponsored Death Squads Integrate 'Opposition Forces'," *Global Research*, May 28, 2012.

146. Charles Lister, *The Free Syrian Army: A Decentralized Insurgent Band*, Brookings Institute Analysis Paper No. 26 (November 2016). www.brookings.edu/wp-content/uploads/2016/11/iwr_20161123_free_syrian_army.pdf; Jay Solomon and Nour Malas, "U.S. Bolsters Ties to Fighters in Syria," *Wall Street Journal*, June 13, 2012; Eric Schmitt, "CIA Said to Aid in Steering Arms to Syrian Opposition," *New York Times*, June 21, 2012.

147. Ken Dilanian and Kevin Monahan, "Obama Nixed CIA Plan That Could Have Stopped ISIS: Officials," *NBC News*, April 2, 2016.

148. Julian Borger and Nick Hopkins, "West Training Syrian Rebels in Jordan," *Guardian*, March 8, 2013; Karen DeYoung, "U.S. Pledges to Double Nonlethal Aid to Syrian Rebels as Opposition Backers Reach Consensus," *Washington Post*, April 20, 2013.

149. Lister, *The Free Syrian Army*; Orlando Crowsoft, "Syrian Sniper: US TOW Missiles Transform CIA-Backed Syria Rebels Into Ace Marksmen in the Fight Against Assad," *International Business Times*, October 30, 2015; Jeremy Bender, "There are a Lot of CIA-Vetted Syrian Rebel Groups Taking It to Assad," *Business Insider*, October 21, 2015.

150. Wikileaks "The Global Intelligence Files." http://wikileaks.org/gifiles/releasedate/2012-03-19 (accessed May 4, 2012); Russia Today,

"WikiLeaked: Ex-Blackwater 'Helps Regime Change' in Syria," March 21, 2012. https://www.rt.com/news/stratfor-syria-regime-change-063/

151. Christopher Blanchard and Amy Belasco, *Train and Equip Program for Syria: Authorities, Funding, and Issues for Congress* (Washington, DC: Congressional Research Service, June 9, 2015. https://fas.org/sgp/crs/natsec/R43727.pdf; US Department of Defense, *Security Cooperation Programs: Fiscal Year 2017* (Washington, DC: US Department of Defense, May 26, 2017). www.discs.dsca.mil/documents/publications/security_cooperation_programs_handbook.pdf?id=170526

152. Barbara Starr, "Top U.S. Commander Completes Day-Long Secret Visit To Syria," *CNN*, May 23, 2016.

153. Barak Barfi, "Ascent of the PYD and the SDF," The Washington Institute for Near East Policy, Research Notes, No. 32 (April 2016) www.washingtoninstitute.org/uploads/Documents/pubs/ResearchNote32-Barfi.pdf

154. US Department of Defense, "Contracts, Press Operations," US Department of Defense, Release Number CR-143-16, July 27, 2016. www.defense.gov/News/Contracts/Contract-View/Article/873473/; Kate Brannen, "Spies-for-Hire Now at War in Syria," *Daily Beast*, September 8, 2016; Paul Shinkman, "War Zone Contractor's Role Under Trump Questioned," *US News*, April 25, 2017; Aram Roston, "Meet The Obscure Company Behind America's Syria Fiasco," *BuzzFeed News*, September 17, 2015.

155. Barfi, "Ascent of the PYD and the SDF"; Lister, *The Free Syrian Army*.

156. Jennifer Percy, "Meet the American Vigilantes that are Fighting ISIS," *New York Times Magazine*, September 30, 2015; Mark Townsend, "Revealed: UK 'Mercenaries' Fighting Islamic State Terrorist Forces in Syria," *Guardian*, November 22, 2014.

157. Representative Tulsi Goddard, *H.R. 608 Stop Arming Terrorists Act*, 115th Congress (2017–18), Congress.gov. www.congress.gov/bill/115th-congress/house-bill/608; James Carden, "Why Does the US Continue to Arm Terrorists in Syria?" *The Nation*, March 2, 2017.

158. Erika Solomon, "U.S. Backed Syrian Militias Launch Offensive Against Last ISIS Stronghold," *Financial Times*, September 9, 2017.

159. CIA, *Libya's Qadhafi: The Challenges to US and Western Interests*, CIA Report, National Intelligence Estimate 36.5-85, March 1985. www.cia.gov/library/readingroom/docs/CIA-RDP08S02113R000100310001-4.pdf

160. Stephen Engelberg, "Reagan Approval Reported on Plan to Weaken Libya," *New York Times*, November 4, 1985; Jeff McDonnell "Libya: Propaganda and Covert Operations," *Counterspy*, 6, no 1 (November 1981–January 1982). www.cia.gov/library/readingroom/docs/CIA-RDP90-00845R0001001 40005-7.pdf

161. Robert Fisk, "America's Secret Plan to Arm Libya's Rebels," *Independent*, March 7, 2011; Mark Hosenball "Exclusive: Obama Authorizes Secret Help for Libya Rebels," *Reuters*, March 31, 2011.

162. Olivier Corten and Vaios Koutroulis, "The Illegality of Military Support to Rebels in the Libyan War: Aspects of *jus contra bellum* and *jus in bello*," *Journal of Conflict and Security Law*, 18, no. 1 (April 1, 2013), 59–93; John

Barry, "America's Secret Libya War: U.S. Spent $1 Billion on Covert Ops," *Daily Beast*, August 30, 2011.

163. Mark Hosenball, "Mercenaries Joining Both Sides in Libya Conflict," *Reuters*, June 2, 2011.

164. See, for example, Christopher S. Chivvis, *Toppling Qaddafi: Libya and the Limits of Liberal Intervention* (Cambridge: Cambridge University Press, 2013).

165. Paul Shinkman, "US Commandos Expand Anti-ISIS War Into Libya," *US News*, January 29, 2016.

166. Eric Schmitt, "Pentagon Has Plan to Cripple ISIS in Libya With Air Barrage," *New York Times*, March 8, 2016.

167. Nick Turse, "Washington Fights Fire with Fire in Libya," *Middle East Eye*, April 16, 2014.

168. Patrick Wintour, "UN Has Set Dangerous Precedent Says Libya's Oil Boss," *Guardian*, August 11, 2016; Adam Nathan, "Militiamen Who Became Libya's Oil Kingpin," *Politico*, August 25, 2016.

169. Emily Wax and Karen DeYoung, "U.S. Secretly Backing Warlords in Somalia," *Washington Post*, May 17, 2006; Mark Mazzetti, Jeffrey Gettlemen, and Eric Schmitt, "In Somalia, U.S. Escalates a Shadow War," *New York Times*, October 16, 2016.

170. Scahill, *Dirty Wars: The World is a Battlefield*, 193, 127–129; David Axe, "New American Ally in Somalia: Butcher Warlord," *Wired*, September 8, 2011.

171. Wax and DeYoung, "U.S. Secretly Backing Warlords in Somalia."

172. Antony Barnett and Patrick Smith, "US Accused of Covert Operations in Somalia," *Guardian*, September 10, 2006.

173. Mazzetti et al., "In Somalia, U.S. Escalates a Shadow War"; P. N. Lyman, "The War on Terrorism in Africa," in John W. Haberson and Donald S. Rothschild, eds., *Africa in World Politics – Reforming Political Order* (London: Westview Press, 2009), 278; Ty McCormick, "Exclusive: U.S. Operates Drones from Secret Bases in Somalia," *Foreign Policy*, July 2015.

174. On US military assistance, see Vincent Morelli, *Ukraine: Current Issues and US Policy* (Washington, DC: Congressional Research Service, January 3, 2017). https://fas.org/sgp/crs/row/RL33460.pdf; On Ukraine's use of mercenaries, see Jack Losh, "Ukraine Turns a Blind Eye to Ultrarightist Militia," *Washington Post*, February 13, 2016.

175. David Levine, "US Congress Quietly Enables Funding for Ukrainian Neo-Nazi-Led Azov Regiment," *Global Research*, February 1, 2016; Alexander Clapp, "Why American Right-Wingers Are Going to War in Ukraine," *Vice News*, June 20, 2016.

176. See McFate, *The Modern Mercenary*, 41.

177. Grandin, *Empire's Workshop*, 235.

Conclusions

1. Ariel Ahram, *Proxy Warriors* (Stanford, CA: Stanford University Press, 2011); Ariel Ahram, "Origins and Persistence of State-Sponsored Militias:

Path Dependent Processes in Third World Military Development," *Journal of Strategic Studies*, 34, no. 4 (2011), 531–556.

2. See Neil Mitchell, Sabine C. Carey, and Christopher K. Butler, "The Impact of Pro-Government Militias on Human Rights Violations," *International Interactions*, 40, no. 5 (2014), 812–836; Govinda Clayton and Andrew Thomson, "Civilianizing Civil Conflict: Civilian Defense Militias and the Logic of Violence in Intrastate Conflict," *International Studies Quarterly*, 60, no. 3, (2016), 499–510.

3. William I. Robinson, *A Theory of Global Capitalism: Production, Class and State in a Transnational World* (Baltimore, MD: Johns Hopkins University Press, 2004); David Harvey, *New Imperialism* (Oxford: Oxford University Press, 2003); David Harvey, *A Brief History of Neoliberalism* (Oxford: Oxford University Press, 2005).

4. One concrete example of this was how the US attempted to prohibit photographs of coffins of US soldiers draped in American flags being sent back from Iraq: Ian Roxborough, "The Ghost of Vietnam: America Confronts the New World Disorder," in Diane Davis and Anthony Pereira, eds., *Irregular Armed Forces and their Role in Politics and State Formation* (Cambridge: Cambridge University Press, 2003).

5. Jeremy Kuzaramov, "Distancing Acts: Private Mercenaries and the War on Terror in American Foreign Policy," *Asia Pacific Journal*, 12, no. 52/1 (2014); Andreas Krieg, "Externalizing the Burden of War: the Obama Doctrine and US Foreign Policy in the Middle East," *International Affairs*, 92, no. 1 (2016), 97–113.

6. According to US policymakers. Whether or not this is actually the case is contentious and a source of controversy amongst specialists on the subject. This is especially the case in light of multiple high-profile cases of fraud and misuse of public money. Regardless, what is important is that US policymakers believe that PMCs offer distinct advantages over the US military in providing certain services. See David Isenberg, *Shadow Force: Private Security Contractors in Iraq* (Westport, CT: Praeger, 2009), 22.

7. Moshe Schwartz, *The Department of Defense's Use of Private Security Contractors in Iraq and Afghanistan: Background, Analysis, and Options for Congress* (Washington, DC: Congressional Research Service, 2010).

8. See Stathis Kalyvas, *The Logic of Violence in Civil War* (Cambridge: Cambridge University Press, 2006, reprinted 2009), 106–110.

9. See, for example, Ruth Jamieson and Kieran McEvoy, "State Crime by Proxy and Judicial Othering," *British Journal of Criminology*, 45 (2005), 514; Kristen McCallion, "War for Sale! Battlefield Contractors in Latin America & the 'Corporatization' of America's War on Drugs," *U. Miami Inter-Am. L. Rev.* 36 (Spring, 2005), 341; Colonel Richard Gross, "Different Worlds: Unacknowledged Special Operations and Covert Action," *US Army War College Report* (May, 2009).

10. Sabrina Siddiqui and Ben Jacobs, "Donald Trump: We Will Stop Racing to Topple Foreign Regimes," *Guardian*, December 7, 2016.

11. On the central importance of minerals, see Mark Landler and James Risen, "Trump Finds Reason for the US to Remain in Afghanistan: Minerals," *New York Times*, July 25, 2017. On US plan to create more militia programs in Afghanistan, see Sune Engel Rasmussen, "UN Concerned by Controversial Plan to Revive Afghan Militias," *Guardian*, November 19, 2017. For US debates over plan to create a private army, see Rosie Gray, "Eric Prince's Plan to Privatize the War in Afghanistan," *The Atlantic*, August 18, 2017.

Index

capitalism, global, 5, 6, 8, 11, 12, 13, 15, 139, 141, 163
  and dominance of US, 16, 17, 28, 65, 69, 94, 116, 118, 126, 140, 148, 163, 166
  and global South, 23, 24–5, 27, 28, 57
  and international organizations, 20, 21, 114
  and neoliberal policies, 93, 105, 139
  see also capital, transnational
Carafano, James Jay, 103–4
Carter, Jimmy, 108
Castaño, Carlos, 131
Castro, Fidel, 45, 46, 48, 59–60, 61, 62
Central America, 58, 96, 133
  see also El Salvador; Guatemala, Nicaragua
Cheney, Dick, 117
Chile, 44, 46, 52, 66
Church Committee intelligence review, 52, 56–7, 91, 95, 102
CIA (Central Intelligence Agency), 1, 4, 11, 37, 86, 91, 120, 128, 130, 153, 155
  airlines contracted by, 55–6, 60–1, 63, 64, 88, 92, 101–4
  and covert operations, 40, 42, 47, 49, 59, 61, 62, 84, 85, 92, 96, 103, 124, 152
  and development of paramilitaries, 79–80, 97, 99, 144–45
  and regime destabilization, 35, 44, 45, 46, 52, 57, 120, 148–49, 152, 160
  and training, 75, 77, 89, 108–11
  and unconventional warfare, 42, 43, 51, 53, 54, 58–60, 158–9
Civil Air Transport (CAT), 42, 55, 58, 59, 60, 61, 62, 87, 88, 103,165
Civilian Armed Force Geographical Unit (CAFGU), 121–22
Civilian Military Assistance (CMA), 109

civilian self-defense forces (CSDF), 65, 67–8, 73–4, 75, 77, 78, 79, 165, 167
  and CADs program, 127
  and CIDG program, 84, 85, 86, 88
  and PACs in Guatemala, 120–1
  strategic mobilization of, 84–7, 119–20, 143–4, 150
civilians, 31, 37–8, 51, 68, 69–70, 89, 107, 130, 134, 135, 161
  drone strike deaths of, 147
  and intelligence gathering, 73–4, 79, 120
  UN mandate to protect, 160
  violence towards, 63, 71, 108, 124, 135, 151, 153–4, 156–57
  see also civilian self-defense forces
Clark Amendment, 95–6, 97
Clemens, Kristi, 156
Cold War, 8, 17, 18, 24, 25, 36, 41, 57–9, 75, 168
  and counterinsurgency, 32, 71, 72, 73, 78, 81, 123, 144
  and covert action, 46–50
  and period following, 4, 5, 104, 114–18, 120–2, 124–8, 136
  policies of, continued, 6, 7, 13, 30–1, 93–4, 107, 113, 119, 137, 139, 142, 144, 148, 162
  pursuit of plausible deniability during, 47, 49, 57, 165
  revised understanding of, 21–2
  and unconventional warfare, 42–6, 51–6, 60–4
Colombia, 3, 66, 81, 82–3, 103, 104, 116, 129–37, 147
communism, 42, 46, 51, 54, 65, 69, 80, 82, 93, 94, 118
  decline of, 11, 114, 116
  and popular revolution, 24, 44
  in South East Asia, 58–9, 62–3, 121
  and US policy of containment, 17, 21, 25, 48, 47, 77, 81
  see also anti-communism
conflict, low intensity, 28, 30, 33, 42, 93, 94–5, 96, 107, 112, 117, 118